Synthetic Biology and Morality

Basic Bioethics
Arthur Caplan, editor
A complete list of the books in the Basic Bioethics series appears at the back of this book.

Synthetic Biology and Morality

Artificial Life and the Bounds of Nature

edited by Gregory E. Kaebnick and Thomas H. Murray

The MIT Press
Cambridge, Massachusetts
London, England

This book was set in Sabon by the MIT Press.

Library of Congress Cataloging-in-Publication Data

Synthetic biology and morality : artificial life and the bounds of nature / edited by Gregory E. Kaebnick and Thomas H. Murray.
pages cm.—(Basic bioethics)
Includes bibliographical references and index.
ISBN 978-0-262-01939-2 (alk. paper)—ISBN 978-0-262-51959-5 (pbk. : alk. paper)
1. Bioethics. 2. Bioengineering—Moral and ethical aspects. 3. Synthetic biology—Moral and ethical aspects. 4. Artificial life—Moral and ethical aspects. I. Kaebnick, Gregory., editor of compilation.
QH332.S97 2013
174.2—dc23
2012049083

Contents

III Values and Public Policy

Series Foreword

Glenn McGee and I developed the Basic Bioethics series and collaborated as series coeditors from 1998 to 2008. In Fall 2008 and Spring 2009 the series was reconstituted, with a new Editorial Board, under my sole editorship. I am pleased to present the thirty-seventh book in the series.

The Basic Bioethics series makes innovative works in bioethics available to a broad audience and introduces seminal scholarly manuscripts, state-of-the-art reference works, and textbooks. Topics engaged include the philosophy of medicine, advancing genetics and biotechnology, end-of-life care, health and social policy, and the empirical study of biomedical life. Interdisciplinary work is encouraged.

Arthur Caplan

Introduction

Gregory E. Kaebnick and Thomas H. Murray

In May 2010, as The Hastings Center was holding the last of three meetings in a project about the ethical issues of a new but still rather obscure branch of genetic technologies that had been dubbed synthetic biology, news began to trickle in that the field was about to break into the national consciousness. Researchers at the J. Craig Venter Institute (JCVI) had successfully managed to synthesize the genome of a bacterium, *Mycoplasma mycoides*, insert it into a cell of a closely related species, *Mycoplasma capricolum*, from which the genome had been removed, and produce what was to all appearances a fully functioning *M. mycoides*. They let the cell carry on with its new life for a while, doing what *M. mycoides* does and producing a few billion *M. mycoides* progeny, to prove that the experiment had worked. The only difference between the new cells—the original and all its progeny—was that the genome included sequences that provided, in code, the names and e-mail addresses of some of the researchers involved in its synthesis, along with some inspirational quotations, including the famous line of Richard Feynman's, "What I cannot create, I do not understand."

Although only the genome had been synthesized, the researchers reasoned that the cell was under the control of the genome, had quickly become the product of its genome, and was therefore a "synthetic cell." And given the genetic differences inserted into the genome, the researchers reasoned that they had, in effect, created a new species. They named it *Mycoplasma mycoides JCVI-1.0*. The name, with its implicit suggestion that the species was just a first draft, gave the promise of more to come.

Press calls began flooding in even before news of the cell was public, and in the ensuing weeks and months, the House Energy and Commerce Committee held hearings on its implications, President Obama charged the Presidential Commission for the Study of Bioethical Issues with delivering a full report on synthetic biology, and the commission held its own series of meetings. The work raises a welter of ethical concerns, none of which are unprecedented, but which arise in synthetic biology in sharp and sometimes perplexing forms. There are questions, for example, about the extent of scientific freedom, the value of open scientific publication, the responsibilities of scientists and government to protect or promote the public good, the desirability (or possibility) of guiding scientific progress and industrial development so as to promote equitable distributions of goods and harms, and the necessity of weighing and comparing potential benefits and harms to society—all familiar enough issues in outline, but sharpened in the case of synthetic biology by the interesting ways in which the technology can be applied.

For there is certainly something new in the outcomes that synthetic biology might turn out to have. On the one hand, synthetic biology has been trumpeted as the enabling technology for a second industrial revolution in which cellular factories obsolete many existing ways of producing goods and in which sugar, the substrate on which cells will operate to produce new goods, replaces oil as the most important industrial commodity. On the other hand, the technology also raises concerns about dangers and deliberate misuses that are reminiscent of concerns nuclear physicists had in the 1940s about their insights into the energy contained in the atom and the possibilities for releasing it. A rogue state or a bioterrorist might recreate smallpox or the 1918–19 influenza strain, or perhaps create new kinds of pathogens. Another, more prosaic set of concerns centers on mere safety, or on what some have called "bioerrorism," to complement the concern about bioterrorism: synthesized organisms might escape from the laboratory or factory, turn out to have properties different from what was intended and predicted, perhaps mutate to acquire them, and become established in the wild, posing a threat to public health, to agriculture, or to the environment.[1] Finally, synthetic biology might turn out to have profound economic consequences. If the field is the beginning of a new industrial age, then some of the changes might be unwanted; one concern is that the cellular factories will run on complex sugars extracted from sugarcane or other plants grown in vast tracts of the developing world, with the effect that food crops needed by some of the world's poorest people might be replaced by crops used to produce materials consumed in wealthier countries.

In some ways, however, the foremost question is about the very idea of creating a "synthetic cell," aside from the possible outcomes. Synthetic biology seems to involve a quest for a degree of control over the basic mechanisms of life that human beings have never attained before; is that quest desirable? Is it troubling? These questions, too, are not without precedents; genetic engineering (which has led to the genetically modified organisms now widespread in agriculture), somatic cell nuclear transfer (or cloning, the technique used to create the sheep Dolly in 1996), and assisted reproductive technologies (such as in vitro fertilization) have raised similar concerns. The power to split the atom also seemed to some to confer godlike powers on mere mortals. Nonetheless, synthetic biology raises questions about the human relationship to the natural world and about human control over living things in an interesting and perplexing form. If there is anything genuinely morally new in synthetic biology, it is this issue.

An Overview of the Technology

The barrage of scholarly reports and media coverage about how synthetic biology will change the world give the impression that the label refers to a specific phenomenon—indeed, the very fact that it has a specific label gives that impression. But in fact, exactly what synthetic biology is is itself debated.

Some commentators describe synthetic biology very broadly as a new way of doing biology: synthetic biology is the attempt to engineer biological constructs—parts, systems, organisms—in order to serve particular human functions. A banner across the top of a website maintained collectively by people working in synthetic biology declares that synthetic biology is just "(a) the design and construction of new biological parts, devices, and systems, and (b) the re-design of existing, natural biological systems for useful purposes."[2] Traditional biology has been "analytic"; it has been about understanding life as it is. "Synthetic" biology is aimed at constructing what might be. A report from the Rathenau Instituut, a Dutch organization that specializes in technology assessment, describes the difference in revolutionary language: "Until recently, biotechnologists focused on modifying the DNA of existing organisms (genetic modification). Synthetic biologists go one step further. They want to design new life and construct this from scratch."[3]

Others in the field, however, are more inclined to emphasize the continuities between contemporary work and much older work. Using biological understanding to construct things is not in itself new, for example. Some trace the start of synthetic biology to 1828, when the German chemist Friedrich Wöhler learned how to make synthetic urea. What's new in synthetic biology is perhaps merely that our understanding of biology has progressed to the point that, combined with ever faster and more powerful tools for manipulating biological systems—in particular, with machines to synthesize medium-length strings of DNA—the possibilities for construction are greater.

In principle, defining synthetic biology merely in terms of building recognizes no limits to the kinds of biological systems or organisms we might work on. Complex, multicellular organisms engineered to do what we want can be envisioned: gourds that grow big enough to be used as dwellings, or rats that sniff out and attack unexploded ordnance. In practice, however, synthetic biology is about the creation or modification of individual cells, and chiefly about the creation or modification of single-celled organisms. So understood, synthetic biology is one way of developing and applying the basic biotechnological skills used to create genetically modified organisms in agriculture. It consists in developing and applying them to comparatively simple organisms—that level of life which scientists stand the best chance of being able to "engineer" and which industrialists stand the best chance of being able to use. In one way, then, synthetic biology is agricultural biotechnology taken to its higher levels—at which we do not merely tinker with systems (sometimes haphazardly, guessing at the possible results) but design rationally from the ground up—but in another way, it is agricultural biotechnology taken down to its rudiments, where it is applied only to the simplest possible organisms.

The work labeled as synthetic biology is advancing along various broad streams,[4] and this is another sense in which the label "synthetic biology" does not refer to a discrete phenomenon. JCVI's announcement that it had synthesized the genome of *Mycoplasma mycoides* falls into a subset of synthetic biology sometimes referred to as "synthetic genomics," which aims at creating complete organismic genomes with the eventual goal of simplifying a genome, that is, stripping out unnecessary genes so that it contains only the genetic material needed to sustain life. The resulting "minimal genome" could, the researchers hope, serve as a kind of chassis or platform that could be augmented with specialized genetic sequences that cause the organism to behave in programmed ways. With the unnecessary genes removed, the theory goes, the organism might prove easier to modify and its behavior might be easier to predict and control. Creation of a fully functioning *M. mycoides* with a synthetic genome was a step toward this goal; it was proof of principle that the fabrication of a living cell with a minimal genome should be possible.

A second stream of work in synthetic biology is what might be called the "BioBricks" or bioparts line, focusing on "DNA-based device construction."[5] This line of work aims at the development of catalogs of standardized, interchangeable genetic sequences that, inserted into a cell's genes, would reliably cause the cell to function in specific, well-characterized ways. Each sequence would code for a specific, limited function, and multiple sequences would be conjoined in longer sets to produce the desired cellular behavior. Ideally, the bricks could be recombined in new ways, like Lego bricks, to produce a different desirable behavior.

In principle, synthetic genomics and bioparts development could be complementary approaches: one could use bioparts to augment a minimal genome. In practice, they are in some tension with each other. Synthetic genomics is led by JCVI and its corporate sponsor, Synthetic Genomics Inc., and aims at developing patented, commercialized products. The development of bioparts is envisioned as a kind of crowd-sourcing approach in which genetic sequences could be obtained literally as off-the-shelf parts from catalogs open to the public. Thus the bioparts stream consists not just of technical advances but also of the development of the institutions and intellectual property law that can help bring the vision to reality.

Protocell development, a third broad stream of work in synthetic biology, is similar to synthetic genomics in that it, too, aims to produce a simplified cellular platform or "chassis." Protocell development starts, however, not with cellular genomes but with the overall set of biological structures and mechanisms necessary for cells: cell walls, intracellular transport, metabolism, and replication. In principle, protocell development might use cellular materials and mechanisms different from those found in any existing organisms. For some, this is the ultimate goal: the creation of "a new form of life," "a genuinely new living entity, . . . one not based on biology and not made out of the customary biological ingredients: no DNA, no conventional biomolecules, no cell membrane of the ordinary type, no nucleus, no mitochondria, no endoplasmic reticulum or any of the other innumerable vital trappings of normal, orthodox biological cells."[6]

The creation of true protocells remains some way in the future, however. The work in synthetic biology that is closest to actual applications—to developing organisms that can be put to industrial or other use—not only makes use of the trappings of existing organisms, but makes use of specific existing organisms. One example is the modification of *Escherichia coli* and later baker's yeast to produce artemisinic acid, from which the antimalarial drug artemisinin can be produced. The modification involves multiple genes from various organisms and the creation of a metabolic pathway that does not exist in nature. The technology was further developed by Amyris Biotechnologies and transferred to the pharmaceutical company Sanofi-Aventis. Another example is work being done on various organisms to develop new ways of producing biofuel. The organisms range from bacteria that could turn complex sugars into fuels to algae that could produce fuel photosynthetically.

Efforts to reengineer existing organisms could be described as advanced genetic engineering as easily as synthetic biology. It consists largely in identifying useful genetic sequences in other organisms and transferring them into the target organism. It differs from traditional genetic engineering chiefly in that it can bring together genes from many different organisms to create integrated genetic modules; the ability to predict what the genetic reshuffling will accomplish is better; and the modules created could turn out to be reusable in other ways—they would be standardized biological sequences, to some degree. But it is not the creation of mix-and-match biological parts, nor the synthesis of entire genomes, nor certainly the creation of nonorthodox forms of life.

In short, the label "synthetic biology" does not refer to any one kind of work and is not sharply differentiable from some other kinds of biotechnology. It usually but not necessarily involves DNA. It might lead to new kinds of organisms—even to a new, "nonorganic" kind of biochemistry—but for the time being, the most promising work centers on making modifications to existing organisms. It might eventually be applicable to complex organisms, but for the time being it centers on the modification of single-cell organisms. It is driven partly by scientific curiosity—it would help us understand how genes work, even how life in general works—but it aims at a wide array of industrial opportunities and other social goals. The field of synthetic biology is similar to the ethical questions it raises: by and large familiar, but taking interesting and sometimes perplexing forms.

Outline of the Book

The question about the very idea of synthesizing organisms, besides being the best candidate for a genuinely new question associated with synthetic biology, is also a threshold question—a question, that is, that is logically prior to the others. It is not dependent on how technology turns out to affect humans; the question is precisely whether the work is morally troubling, or attractive, or morally neutral—whatever the consequences for human well-being.

This question was at the heart of a research project conducted at The Hastings Center around the time JCVI unveiled its "synthetic cell." The goal of the Hastings project was less to answer the question than to support its articulation and examination. In keeping with that goal, the collection of essays in this volume provides a range of very different views.

The first three essays focus on the basic question about the ethics of making new organisms. In the opening essay, Andrew Lustig engages in a little "conceptual cartography": he reviews "the complex and often conflicting philosophical, thematic, and sociocultural meanings associated with appeals to nature and the natural," then tries to articulate the religious and secular concerns that appeals to nature seek to capture. He closes with some thoughts about the relevance of these appeals to policy making and public discourse. This last issue was important throughout the project and recurs in many of the essays. Thus Lustig's essay both launches the moral analysis and captures the overall sweep of the project.

In the second chapter, Joachim Boldt develops the position that synthetic biology is deeply troubling in and of itself—intrinsically troubling. "Synthetic biology is characterized," he writes, "both by a perspective of creation and a reductionist notion of life": it arguably makes human beings capable for the first time in human history of creating life, yet it imposes a conceptual framework within which living things are objects only, not subjects. There is a moral loss in this transformation, even if synthetic biology is restricted to the synthesis of single-cell organisms.

The third chapter takes the converse view. Gregory Kaebnick proposes that there are a variety of different ways of articulating the intrinsic concern synthetic biology poses, but that most of these will be unpersuasive to most people. The most compelling way of expressing the concern is to see it as connected to the ground-level concern in environmental ethics about human alteration of the natural world. How far should humans go in altering nature to provide human benefits; what is the line between altering nature and accommodating nature? But Kaebnick proposes that in this way of understanding the concern, the fact that synthetic biology is limited to the synthesis of microorganisms—and to serve ends that already require extensive alteration of nature—is indeed significant.

The next three chapters broaden the focus to ask whether synthetic organisms might, far from being morally troubling, actually be valuable in and of themselves. In chapter 4, Mark Bedau and Ben Larson deploy work in environmental ethics, including a critique of synthetic biology developed by Christopher Preston several years ago, to argue that creating synthetic organisms should not be seen as violating the intrinsic value of nature. In fact, claim Bedau and Larson, a broad understanding of the intrinsic values that environmentalists find in nature could actually apply to synthetic organisms as well—thus, far from giving reason to refrain from synthesizing organisms, environmental ethics puts us in a position to find value in those organisms. They possess the same kinds of basic properties that naturally occurring organisms possess, excepting "the value of wilderness and the value of a natural evolutionary history." Even those values, however, could be acquired. "If synthetic life-forms were released into the environment and left to fend for themselves, they could adapt, mutate, and evolve in directions unanticipated today. . . . So, the descendants of synthetic life-forms could start to become wild and start to accumulate a complex contingent evolutionary history." The problem with synthetic biology, Bedau and Larson suggest, is a matter of culture shock. "Synthetic biology disrupts our picture of our place in the cosmos." It gives us, just as Boldt suggested, a different view of what life fundamentally is. "To many people, the mechanistic implications of synthetic biology deflate human freedom and morality."

In chapter 5, John Basl and Ronald Sandler argue that synthetic biology generates a series of moral puzzles. The basic problem is this: Organisms are widely held to have what Basl and Sandler call "inherent worth," meaning that they have value in themselves, and indeed that their interests ought to be borne in mind by any moral agent; artifacts, however, are widely thought to have only instrumental value. "What, then, are we to make of synthetic organisms . . . ? Should they be regarded as possessing inherent worth or not?" Basl and Sandler conclude that synthetic organisms apparently must have inherent worth, at least if nonsentient natural organisms do. The conclusion poses no practical problems for now, since whatever moral status nonsentient organisms have, it cannot amount to much, but if synthetic biology progresses to the point that complex organisms can be created, their moral status might well present objections to the science.

In chapter 6, Christopher Preston proposes that environmental ethics may be essentially neutral with respect to synthetic biology, at least in its limited, microbial form. In an earlier paper, Preston had argued that synthetic biology amounted to "a line in the sand," beyond which the environmentalist's fundamental distinction between the natural and the artifactual was lost forever. In this paper, then, Preston pulls back from this view. On the other hand, he rejects the position Sandler and Basl and Bedau and Larson stake out—that synthetic organisms might actually acquire some of the same values that environmentalists find in other organisms. One reason is that synthetic organisms will still differ from naturally evolved organisms. The other is simply that environmentalists have mostly never attributed inherent value to bacteria. "If the extent to which naturally occurring bacteria have inherent value is zero, then [perhaps] the inherent value of synthetic bacteria is zero."

The final set of chapters bring sustained attention to the issue of whether and how intrinsic moral objections to synthetic biology might be integrated into policy-making and public discourse. In chapter 7, Jon Mandle sets out what is, for many political philosophers, more or less the received view on this question—namely, a Rawlsian account of the so-called principle of neutrality—and then seeks to apply that view to the kinds of intrinsic concerns that surface in discussions of synthetic biology. The Rawlsian view, in a nutshell, is that the state should seek to avoid taking sides on religious, metaphysical, and moral questions that are not absolutely vital to the just functioning of the liberal society. Mandle concludes that the state should refrain from enacting policy that is based on the intrinsic objections to synthetic biology.

Chapter 8, by Bruce Jennings, develops a contrary view. Synthetic biology, and biotechnology generally, "convey civic and moral lessons," writes Jennings, and in deep ways they are the wrong lessons. "Not wrong necessarily in terms of the direct effects of the science within its still limited biological and operational range—I have nothing against growing microorganisms that can serve as clean fuel, or that can mitigate the environmental damage of oil spills—but wrong in terms of the indirect influence they have." Synthetic biology teaches us to accept "the human appropriation, manipulation, and 'engineering' of nature and of life." What is needed, Jennings holds, is to shift precisely in the opposite direction, away from this vision, toward a discourse that promotes the accommodation of nature and limits on human behavior. And given the importance of these concerns for the environment and for human life, they are certainly suitable bases for public policy. The principle of neutrality "is totally inadequate for capturing the significance of the concerns I am trying to articulate here."

The volume concludes with an extended sociological analysis, by John Evans, of religious and philosophical objections to biotechnology, how they have fared so far in bioethics, how they ought to figure in public policy making, and how they are likely to carry over into the debate about synthetic biology. Evans argues that biotechnology can convey new understandings about what it means to be human, and that an objection to a new understanding can be an acceptable basis for public policy-making, if most of the public share the objection. But he is skeptical that synthetic biology would impart a new view of what it means to be human. Evans identifies two possible views, mirroring the lessons that Boldt develops: synthetic biology might teach us we are creators, not just manipulators, and it might teach us we are only biological machines. In short, it might teach us that we are like gods, or that we are entirely un-godlike.

Evans argues, drawing on sociological research he has conducted on views about other biotechnologies, that neither lesson is likely actually to raise problems. Synthetic biology will not teach us that we are creators, he says, because we already see ourselves as creators in the relevant respects. And it is unlikely to teach us that we are just biological machines, he believes, because people tend to see human beings as morally and metaphysically different from microbes. We may well come to think of microbes as biological machines—indeed maybe we already do—but we are unlikely to draw the conclusion that humans must also be machines. Thus Evans defends objections to biotechnology in general, but finds objections specifically to synthetic biology wanting—again, because of the current limited, microbial state of the technology.

Each of the three sets thus illustrates a movement toward a middling and nuanced position about the human relationship to nature. The project aimed to articulate a range of views rather than to reach consensus, and overall, the chapters differ with each other sharply. There is a wide gap between Boldt, on the one hand, and Bedau and Larson, on the other. The chapters by Lustig, Kaebnick, Preston, and Evans, however, all seek to work out the possibility that a general concern about nature would not necessarily translate into opposition to synthetic biology. In fact, resisting human alteration of nature need not translate into opposition to synthetic biology. It is possible, instead, to draw important distinctions between different kinds of living things or different ways of applying synthetic biology, or to trade off different kinds of human alterations of nature.

By contrast, those who objected to synthetic biology were more impressed by conceptual commonalities within the natural world, among living things, and across science. To them, synthetic biology reflects a stance, evident in other biotechnologies as well, that will miss something important about life, or send us in the wrong direction with respect to nature. It will promote a discourse, Jennings holds, that promotes the alteration of nature for human ends, and though particular applications appear unobjectionable when considered in themselves, synthetic biology as a whole belongs to a broader trend that is morally and environmentally damaging and ought in some way to be opposed. Boldt puts his concerns about synthetic biology in terms of "the ontological characteristics of life." He argues that "life is always, albeit to differing degrees, in possession of a capacity of self-renewal, exchange, and freedom," and that synthetic biology, which tries to reduce organisms to their constituent parts instead of understanding them as wholes, "will necessarily miss this part of the reality of life."

Major new developments in biotechnology have sometimes led to efforts to create subfields of bioethics. "Genethics" was coined to address the ethical issues of genetic science, "nanoethics" for those of nanotechnology, and "neuroethics" for those of neuroscience. Predictably, then, the emergence of synthetic biology has led to calls for "synbioethics" or "synthethics." This book treats questions about the intrinsic morality of synthesizing organisms as versions of questions that have been around for a long time, that previous biotechnologies also raise, and that therefore do not constitute a new ethical inquiry. Those questions take an interesting and illuminating form in synthetic biology, however, as perhaps, do other questions that synthetic biology raises—questions about the intrinsic value of knowledge, about how to balance risks against potential benefits, about liberty, and about justice. Even if inquiry about those questions is not new territory, then, it nonetheless breaks new moral ground.

Notes

1. D. Caruso, *Synthetic Biology: An Overview and Recommendations for Anticipating and Addressing Emerging Risks* (Washington, DC: Center for American Progress, 2008).

2. http://syntheticbiology.org/.

3. Rathenau Instituut, *Constructing Life: The World of Synthetic Biology* (The Hague: Rathenau Instituut, 2007), 2.

4. There are other ways of dividing up the field, however. Bottom-up versus top-down, for example.

5. Maureen A. O'Malley, Alexander Powell, Jonathan F. Davies, and Jane Calvert, *BioEssays* 30, no. 1 (2008): 57–65.

6. Ed Regis, *What Is Life?* (New York: Farrar, Straus, and Giroux, 2008), 3.

I

The Human Relationship to Nature

1

Appeals to Nature and the Natural in Debates about Synthetic Biology

Andrew Lustig

Synthetic biology is the field whose aim is "to extend or modify the behavior of organisms and engineer them to perform new tasks" (Andrianantoandro et al. 2006).[1] Descriptions of the field vary, of course, but generally, synthetic biology involves one of two approaches: building living organisms from raw materials, and building organisms from parts already found in living organisms. The first "uses unnatural molecules to reproduce emergent behaviors from natural biology, with the goal of creating artificial life," and the second "seeks interchangeable parts from natural biology to assemble into systems that function unnaturally" (Benner and Sismour 2005).

Both of these approaches include either explicit or implicit explicit appeals to concepts of nature and the natural. But whether such terms and distinctions clarify or obfuscate matters depends on one's vantage. On the one hand, the aim to engineer "unnatural" organisms or molecules may seem continuous with uncontroversial forms of biotechnology and other human interventions into nature. After all, biotechnology, broadly defined as the general effort to employ, shape, and redirect living materials for specific human purposes, is as old as fermentation, as ordinary animal breeding and human agriculture. On the other hand, according to some critics, it is different from other techniques in kind rather than merely degree because it involves novel efforts at synthesizing living materials "from the ground up" for specific human aims. On this reading, synthetic biology (especially the first of the approaches described above) moves beyond merely revising or recombining natural givens and involves a basic reorientation toward living organisms. This form of engineering blurs conceptual and normative lines between organism and artifact, between organic and synthetic, between living and nonliving.

The issues raised by such criticism deserve to be placed in broader context, because they involve judgments about the relevance of appeals to nature in bioethics and the life sciences more broadly, as well as in ecology and environmental studies. Virtually all commentators agree that something, however nebulous, is at stake in such appeals to nature; indeed, perhaps paradoxically, the existential depth and power of such appeals may be reflected in the fact that they are so often raised rather inchoately. At the same time, the projects underscore the complex interplay of perspectives that shape notions of nature and human nature, making any reliance on such appeals in public policy deliberations at best difficult, and at worst impracticable.

In assessing synthetic biology, then, we must start by considering whether appeals to nature raise the sorts of intrinsic concerns that are relevant to public policy debates. By "intrinsic" concerns, I mean those moral issues raised about current and prospective developments that resist specification according to the quantifiable factors in risk-benefit analysis that characterize most policy deliberations. From the outset, it is important to acknowledge the political context of democratic pluralism within which debates about scientific policy proceed. That context enjoins two important requirements upon policy makers. First, policy deliberations should seek to engage the full range of values and attitudes that serve both to justify and to constrain policy choices in a manner that respects core differences between and among individuals and communities in their evaluations of particular issues. Second, in light of what is generally called the "liberal neutrality" thesis, particular policy choices on issues like synthetic biology—whether to ban, to regulate, to remain neutral about, or to actively support such developments—must be articulated and justified on the basis of broadly shared "public reasons," rather than relying for their acceptability on more comprehensive but epistemologically privileged appeals. Given these requirements, the central question, for our purposes, is this: What, if any, function should such appeals have for policy discourse and formation apart from, or in addition to, the usual issues of experimental design and oversight, concerns about biosafety and environmental harm, and matters of justice and fairness in allocation of resources and distribution of goods, all of which can, at least in principle, be quantified according to the usual canons of risk and benefit?

I will seek in this chapter to lay some groundwork for the rest of the volume by doing two things at once: pointing out some of the deep problems with appeals to nature, yet also underscoring how deeply ingrained views about nature are in moral thought, indeed in larger worldviews. In the first section, I engage in a bit of conceptual cartography by reviewing the complex and often conflicting philosophical, thematic, and sociocultural meanings associated with appeals to nature and the natural. As we will see, nature is a polyvalent concept. The complex interplay between and among different understandings of nature and the natural makes it often difficult to determine which perspective (or amalgam of perspectives) best captures the attitudes and values that underlie its invocation. In the second section I review a number of religious concerns voiced in appeals to nature and the natural in debates about biotechnology, as well as their secular analogs. Such concerns are often subsumed under the rubric of "playing God." For all its imprecision, that phrase serves to signal deeply felt concerns; as Peter Dabrock observes, "it identifies particularly relevant issues, questions of ultimate concern, which must be handled with due respect" (Dabrock 2009). Moreover, despite its vagueness, I will suggest that it can be analytically sharpened in ways that simultaneously clarify and problematize its meaning. It can help to highlight the importance of attending to a particular religious tradition's understanding of the appropriate relations between God's sovereignty and the legitimate domain of human responsibility toward created nature. Yet because such judgments about the scope of divine authority and human responsibility tend to (or perhaps necessarily) incorporate the same complexity of appeals to "nature," understandings of created nature are likely to be subject to the same difficulties of definition and specification that beset secular perspectives.

In the chapter's final section, I offer some observations about the relevance of intrinsic moral concerns, especially appeals to nature and the natural, to the processes of policy deliberation and formation. In those judgments, which are admittedly tentative, I conclude that, despite differences between earlier forms of biotechnology and what is practiced and proposed in synthetic biology, appeals to nature and the natural function quite similarly in both cases. As a result, the claimed differences between synthetic biology and predecessor biotechnologies emerge, at most, as differences in degree rather than in kind.

The Complexity of Appeals to Nature

Dynamic "Visions" of Nature

Concepts of nature and the natural are, perhaps paradoxically, some of the most widely yet inchoately invoked terms in ethical and policy discussions of biotechnology. Their currency and their inchoateness are perhaps of a piece, because the American public is deluged 24/7 with headlines that trumpet both the promise and the perils of current and prospective developments in biotechnology. As a result, appeals to nature, often implicitly rather than explicitly, rely on notions drawn from different substantive and methodological frames of reference, including (but not limited to) issues formally engaged by philosophers and theologians as matters of ontology, epistemology, ethics, systematic theology, or theological anthropology.

In a broad yet useful organizing typology, James Proctor, a professor of environmental studies, sets forth five perspectives or visions of nature: nature as evolutionary, nature as emergent, nature as malleable, nature as culture, and nature as sacred. As Proctor observes, "The first two of these visions have arisen in the physical, life, and behavioral sciences; the final two have arisen in the social sciences, the humanities, and theology, with malleable nature straddling the sciences and the humanities" (Proctor 2009a). Obviously, any broad typology emerges as, at best, only a heuristic; there are countless other conceptual maps that could be drawn. One advantage to Proctor's discussion, however, is that it employs the language of "visions" rather than concepts, thereby allowing us to recognize the complex admixture of facts and beliefs, description and prescription, sentiment and intuition, which are invariably present in appeals to nature, to human nature, and to the putative links between them. To acknowledge that mixture of modes and motives is to appreciate the difficulty of separating the strands of exhortation, argument, and worldview often interwoven in such discussions.

Another advantage of Proctor's typology is that it allows various interpretations of nature to be loosely plotted on a theoretical and practical continuum, so that different visions of nature may be understood in dynamic or dialectical terms, with a particular perspective having implications for one or more of the others in something of a corrective fashion.

Let me rehearse several examples of that interpretive synergy. Viewing nature as evolutionary, emergent, or sacred is to acknowledge its status and intrinsic powers as in some sense already given, prior to human interactions with it. These visions find both philosophical and religious warrants. From the perspective of critical realism, human concepts of nature and cosmos do not exhaust the meaning or ontological standing of what exists independent of human will, mastery, and resource extraction. Cosmology is a paradigmatic example of the way that nature, writ large, is "wondrous to behold." The experience of wilderness, a la John Muir, can be, as William James suggested, the context for natural mysticism (James 1982). To appreciate nature as that which is nonhuman is to be reminded both of our stance as valuers and of the nonhuman aspects of the world that remain, in vital ways, sources of both mystery and mastery. This sense of nature as a given, apart from human aims and purposes, has had special salience in much of the recent literature of environmental ethics, especially in the writings of Holmes Rolston (Rolston 1988).

In the context of biology, however, this sense of nature's "externality" assumes that we can metaphysically distinguish aspects of nature that are separate from ourselves from those that are "constitutive" of us. Yet even a cursory reading of various episodes in the history of science illustrates what, with only modest reflection, should be obvious. Even a putatively external nature, as a descriptive generalization or as an organizing perspective or vision, is invariably contextualized. Any naive realism about nature as a given would appear to be as misguided as claims for an uncritical realism more generally.

The historian of science David Livingstone argues that the often profound differences between alternative perspectives on nature result from the distinctive ways that particular visions are appropriated and contextualized in different settings (Livingstone 2009). Moreover, cultural readings of nature also shape, and are shaped by, alternative perspectives on space itself, as studies of the meanings of "landscape" help to clarify (Henderson 2009). Even in the literature of ecology, where one might assume that concepts of nature and the natural approach something closer to a consensus interpretation, deep divisions on quite fundamental matters persist among theorists. As Proctor points out, the term "environment," like its analog "nature," is interpreted in different ways for a range of varied and often conflicting purposes (Proctor 2009b). One especially prominent example of such a flashpoint concerns Aldo Leopold's fundamental description of the relative stability of ecosystems (Agar et al. 2008). There are also ongoing debates about the status and permeability of "species" as "natural" kinds, with differing implications for subsequent judgments about the novelty or distinctiveness of human efforts at genetic alteration. Consider, too, Proctor's vision of nature as sacred. Nature as sacred (or as God's creation) will incorporate varying judgments about divine purposes and human attitudes toward the nonhuman world. In the book of Genesis, for example, one finds the roots for an ethic of stewardship, but a stewardship that includes activities of both preservation and cultivation, with clear potential, from Eden onward, for tension in drawing lines between appropriate and inappropriate human responsibilities. Sacred nature also includes quasi-religious attitudes toward nature that, while shorn of the richer worldview that inspired them, remain deep influences on much current spirituality and in the literature of deep ecology.

I have argued elsewhere that the interpretation of nature as *malleable* is perhaps the most fruitful vision for bioethics (Lustig 2009). Viewing nature as malleable has several theoretical and practical merits. Despite its apparent vagueness, it provides something of a conceptual midpoint, a place of intersection for attitudes and values that function more narrowly and exclusively within other perspectives. Moreover, it does so without necessarily incorporating the descriptive or normative priorities often evident in alternative visions. Put more pointedly, the description of nature as malleable brings to the foreground the explicitly normative dimensions of appeals to nature that are otherwise often assumed, rather than argued for. By making explicit these normative dimensions—especially regarding the scope of human responsibility in altering or reshaping nature as "given"—it seeks either to reinforce or to challenge the relevance of other visions of nature when they are invoked as a justificatory or constraining norm.

"Modes of Discourse" about Nature

A few years ago, I was privileged to participate in a project entitled "Altering Nature: How Religious Traditions Assess the New Biotechnologies," sponsored by the Ford Foundation.[2] In that research, five scholarly groups explored the rich history of concepts of nature from various disciplinary perspectives—religious studies, philosophy, science and medicine, law and economics, and aesthetics—leading to a more finely grained interdisciplinary analysis of the sometimes conflicting, but sometimes overlapping, perspectives on nature that can be found within and across the "home pages" of various disciplines. Several conclusions from that project shed light on the intrinsic concerns voiced in discussions of synthetic biology.

First, to conduct a comparative analysis of various interpretations of nature in different disciplines and modes of discourse, whether for fundamental scholarship or for policy analysis, it is vital to develop a common agenda. In the Ford project, participating scholars adopted a framework of categories and questions that provided the template for robust conversation and research. This framework included broadly ontological, epistemological, moral, and aesthetic concerns. For example, different ontological claims about nature—as fundamentally ordered or broken, as random or suffused throughout with telos—result in different judgments about the appropriate scope of human responsibility vis-à-vis the natural world, and about the relations—if any—between natural teleology and human aims. Central questions of fundamental epistemology included judgments about the accessibility of nature to human reason and its status as a trustworthy source of insight. Central moral questions included the relations between humankind and the larger world (including animals, plants, and environmental niches), as well as economic and legal questions concerning the appropriateness of construing nature as property or as a commodity. Finally, aesthetic concerns included the ways that nature is depicted in the literary and visual arts; is nature a Keatsian source of beauty, goodness, and morality or a source of entropy and chaos that must be ordered by human artistic choice?

A second general point. It was quickly evident that, despite a common list of such questions and concerns, the implications of appeals to nature and the natural in biotechnology discussions would not be univocal. Scholars exploring concepts of nature in science and medicine, as well as those analyzing philosophical perspectives, focused greater attention on how particular accounts of nature in their disciplines reflect diverse historical understandings. Scholars exploring religious and spiritual traditions sought to identify patterns of reasoning about nature that serve to distinguish particular traditions in characteristic ways. The legal-economic group combined a rich historical overview of interpretations of nature as property with a careful analysis of genetic patenting in different traditions. It became abundantly clear that appeals to nature and the natural are difficult to analyze with precision for several reasons. As Proctor's language of visions suggests, one's perspective on nature often incorporates a welter of concerns that are difficult to isolate and analyze separately, especially when they function largely at the level of intuition and sentiment. Moreover, judgments about the cogency of such appeals have to attend to the rich variety of historical and cross-traditional interpretations. Particular perspectives on nature and the natural, although invoked as putatively obvious, are often in tension with one another both across and *within* traditions.

A third point. Efforts to specify the relevance of nature as a moral appeal are complicated by the philosophical challenge traditionally labeled the "is/ought problem"; namely, that moral conclusions cannot be drawn merely from descriptive claims about the world. In traditions that are understood as ways of life, the activities of description and evaluation defy easy separation. From the time of the Stoics (if not before), nature has been invoked as the origin of a universal order that justifies moral choices and actions (Lovin and Reynolds 1985, 2). However, developments in both philosophy and science have called into question the conflation of the empirical and the moral, as captured by David Hume's famous criticism of arguments that move from description to prescription without clear warrant (Hume 1973). (Hume was not in fact prohibiting the conclusions, but only criticizing the failure to provide the necessary bridge between empirical facts and moral directives.) On the one hand, in moral philosophy, efforts in the wake of Kant to formalize the logic of moral claims seem to undercut ascriptions of moral value to phenomenal nature. On the other hand, recent work in the history and philosophy of science underscores the vital role of values and value-laden presuppositions in prompting and shaping scientific inquiry. There are efforts, of course, to retain appeals to nature as the empirical basis of morality, but these attempts have generally been seen as an illicit reduction of normative ethics to descriptive ethics. Moreover, on any such account, the concept of nature remains ambiguous. For example, the natural precedents for both competition and reciprocal altruism can be found in nature. Which should become our moral guide? In deciding, we will necessarily appeal to a criterion that transcends nature. "The choice between these attitudes is a moral choice," writes Mary Midgley, "so it cannot be determined [solely] by science" (Midgley 2003, 24).

The Social Embeddedness of Concepts of Nature

Concepts of nature and the natural are critically shaped by social, cultural, and institutional factors. Recent work in critical studies has helped to illuminate how accounts of nature have served to reinforce social inequalities in ways pernicious to the interests of women and other marginalized members of society. Epistemologists, for example, have explored the ways that concepts and methods in philosophy of science are being expanded or recast in light of feminist concerns. Western culture has often equated rationality with masculinity and emotionality with femininity, with the result that traditional epistemology has tended to valorize male forms of cognition (Tuomela 1995, 263). Moreover, a number of feminist philosophers of science have challenged the status of scientific objectivity as defined by the Baconian paradigm that framed the mechanistic philosophy of the scientific revolution. On their reading, especially since the Enlightenment, the scientific method has been centrally linked to notions of prediction, control, and domination of nature. By contrast, these feminists seek to redescribe scientific success as the ability of scientists to heed nature's self-revelations.[3] The shift of metaphor here—from power and domination to attentive and respectful listening—appears to suggest a different understanding of scientific objectivity itself, and of the methods that may be most appropriate for promoting scientific discovery.

Other scholars have made identifying and rectifying persisting gender-based inequalities their central focus. Some historians have argued that the status of women has been largely determined by social structures and practices, especially practices related to reproduction and child rearing. Other feminists have explored the ways that some prominent liberal theorists have insisted on a biologism of innate differences between men and women in order to justify social inequalities in a manner amenable to their own political predilections. Indeed, a significant portion of anatomical research in nineteenth century institution-based science therefore focused on the putatively natural bases of inequality between men and women (and between Caucasians and other races). In the historical efforts of a male-dominated scientific establishment to identify natural differences in anatomy and biology as the basis of social inequality, the privileged place of white males in society and the professions was justified in light of those putative differences. In that division of labor, a larger working parallel also emerged: "Nature is to female as culture is to male" (Ortner 1972). The latter distinction has emerged as a basic construct in a number of disciplines, including literature, anthropology, and history, which have been subject to major critiques by feminists on that basis.

Carolyn Merchant has elaborated this parallel by exploring the depictions of nature as female in the "gendering" of nonhuman nature (Merchant 1980). "Nature as mother" emphasizes aspects of emotionality, nurture, and feeling, and was, according to Merchant, central to earlier organic conceptions of nature. A second, contrasting image, however, identified a female nature as the source of chaos and disorder; again, on Merchant's reading, that image became increasingly important in the context of the scientific revolution, wherein science was viewed as the appropriate extension of human power over an unruly nature. The political implications of such emphases have been central to recent academic discussions among feminists, especially in the "hermeneutic of suspicion" they direct against patriarchal claims that link social inequalities to differences in biology, and in their exploration of gender as a social construct or ideological projection. At the same time, ecofeminists have sought to critically retrieve an earlier vision of a more "organic" nature as a necessary and corrective element in articulating an ethic of environmental responsibility (Gaard 1993; Merchant 1995; Ruether 1992).

The contributions by scholars in critical studies, then, sensitize us to the broader historical, social, and cultural factors that invariably shape particular interpretations of nature and the natural, especially their putatively normative implications. In the context of postmodernism, we are regularly reminded that there are no such things as "naked" facts, that all epistemological and normative claims are contextualized. What is true of "facts" is even more obviously true of concepts. In assessing the cogency of appeals to nature, recent scholarship reminds us to foreground certain key questions that may have been overlooked by earlier, more armchair approaches to intellectual history. For example, who has "constructed" the idea of "nature"—which persons, professions, or social classes are its primary "authors"? Toward what audience is the concept directed and for what purposes? Who benefits from or is privileged by a particular account or interpretation? In what ways does the concept function as a source and/or product of social relations? Do those at the margins of the institution or tradition within which the concept is developed agree with its articulation and implications, or would they describe nature in different terms? Such questions, and our efforts to answer them, necessarily add a further layer of complexity to the conversation.

Religious Concerns about Biotechnology: The Language of "Playing God"

Appeals to nature, often articulated via the metaphor of "playing God," have been invoked at key moments in the history of debates between science and religion. In recent decades, the playing God metaphor has featured prominently in bioethics debates on topics ranging from assisted reproduction to end-of-life decision making; it has also been invoked in reaction to various developments in biotechnology, particularly as a way to articulate concerns raised by genetic engineering. The metaphor has salience and currency both in expressly religious discussions and in certain secular perspectives that bear family resemblances to the religious perspectives.

As with all metaphors, the metaphor "playing God" serves to illuminate some features even as it necessarily obfuscates other aspects of the issues under scrutiny. As brought to debates on particular biotechnologies, including techniques of synthetic biology, "playing God" is meant to capture three sorts of religious concerns. First, it is often invoked in deontological fashion to express the notion that certain borders or boundaries should not be breached because to do so constitutes a transgression or violation of a domain or realm deemed "off limits" to human action and reserved, if you will, as God's domain. Second, "playing God" may express judgments about certain attitudes, values, or virtues that are appropriate responses to created nature as God's own, including attitudes of awe, wonder, and humility, with a corresponding respect of and deference to natural boundaries. Third, and perhaps more problematically, "playing God" is meant to suggest some straightforward, even isomorphic, correspondence between the ordering of the world, identified with notions such as fixed species or natural kinds more broadly, and the creative will and purposes of God.[4]

Consider each of these meanings in turn. First, the notion of "playing God" as expressing a transgression or illicit boundary-crossing is one familiar to religious ethicists, who have been largely critical of its cogency as a blanket negative judgment. The objections to the notion of illicit transgression or boundary-crossing are of several sorts. On the one hand, religious assessments of the relations between a putative order of nature and the purposes of a Creator God vary widely both across and within religious traditions. To be sure, the Abrahamic faiths all share core convictions about the sovereignty of God, but that theme is linked to varying accounts of human responsibility, appropriate activity, and stewardship with respect to the created order. Moreover, the status of the created order as accessible to human reason and moral insight varies widely, with differing implications, from fairly conservative to fairly liberal, about the appropriateness of human efforts to alter, amend, or emend natural "givens." Much of traditional theology, especially but not limited to natural law reasoning in Roman Catholicism, was formulated in a premodern context, informed by a hierarchical and fairly static metaphysics not easily reconciled with the historicism at work in evolutionary accounts of nature and human nature. Consequently, the correspondences to be drawn between natural phenomena or states of affairs and God's creative or ordaining will face challenges that are *both* scientific and theological. Classical scientific accounts of a "great chain of being," inspired by neo-Platonic sources, cannot be reconciled with neo-Darwinian understandings of evolution and speciation. Theologically, the implausibility in the postmodern context of essentialist accounts of natural kinds and human nature make it difficult to identify unambiguous transgressions of nature; there are no longer any clear or fixed lines.

The difficulty of drawing clear lines sufficient to specify the illicit boundary-crossing is exacerbated by certain fundamental attitudes at work in different perspectives in theological anthropology. One might see these reflected by the two imperatives somewhat in tension in Genesis, wherein God places Adam and Eve in the Garden of Eden and enjoins them "to work it and take care of it" (Gen. 2:15). Depending on how one interprets that joint command, the ambit of appropriate human stewardship will be drawn more narrowly or more broadly. A more conservative reading of the effect of the Fall—both on nature and on human capacities to know the good and to act upon it—will tend to limit human responsibility to acts of "taking care" of the created order, largely by "conserving" it. A more expansive reading of the Genesis injunction will tend to justify a far greater range of appropriate human responsibility. For example, interpretations of the latter sort tend to employ the language of human beings as "created co-creators" with God. Or again, among Jewish commentators, there may be an emphasis on the notion of *tikkun olam,* with a responsibility to repair or restore a creation which has been broken from the first (Zoloth 2008). From the latter perspectives, drawing boundaries with sufficient precision to distinguish licit from illicit interventions will prove difficult on two grounds: the language of co-creation is less likely to interpret current natural processes as expressions of God's will and far more likely to affirm the appropriateness of human partnership in restoring or even enhancing the natural world.

Nonetheless, one would expect to find that appeals to *created* nature will tend to reframe, perhaps even transform, philosophical or scientific accounts of nature as an independent domain, because construing nature as creation situates judgments about the morality of human interventions into the natural world within a theological context. Across the range of religious perspectives on biotechnologies, including synthetic biology, one should expect to find at work descriptions of God's creative purposes, the effects of sin on the natural order, the availability of created nature as a source of normative insight, and the scope of human responsibility vis-à-vis the natural order.

Consider the range of Christian perspectives on created nature, all of which are represented in recent theological discussions and all of which are of general relevance to assessments of biotechnological developments, including those in synthetic biology. In one perspective, nature is viewed as intrinsically valuable; as Genesis 1:31 says, "Then God looked over all he had made, and he saw that it was very good." This perspective will likely be linked to a primary emphasis on humans as stewards of created nature according to God's initial purposes. From a different vantage, nature may be viewed as essentially disordered after the Fall as the result of human sin, with differing implications for assessments of the appropriateness of its alteration by humans. In light of the deleterious effects of sin on both nature and human reason, this perspective may call for humility and caution in our dealings with nature. Alternatively, accounts that find nature a relatively trustworthy source of insight may emphasize the appropriateness of human responsibility to work as partners with God's purposes. For example, in traditional Roman Catholic moral theology, reasoning about the appropriate pursuit of human goods establishes a set of natural-law-based duties that remain transparent to human reasoning after the Fall. Classical Protestant thought has been less prone to appeal to a fallen nature as an unambiguous source of moral insight, but a number of recent Protestant discussions of biotechnology have emphasized a central and positive role for human co-creativity in repairing, restoring, and even enhancing the created order (Peters 1995).

A second, largely cautionary, sense of "playing God" is meant to suggest attitudes and dispositions that are appropriate to human engagements with nature. For example, among the worldviews of many indigenous peoples, nature is viewed as the intrinsic expression or repository of the sacred (an understanding that may be tied to animism, emanationism, and visions of a larger natural or cosmic harmony). It is difficult to offer broad generalizations about such worldviews, but they include a shared sense of an appropriate place for beings, including humans, within a larger frame of reference interpreted in spiritual terms. The language of appropriate dispositions and virtues with respect to nature is also evident in the Abrahamic traditions. While Judaism, Christianity, and Islam clearly distinguish God from nature, they view nature as God's good creation, with its own standing and worth that are not reducible to human ambition and will. Gerald McKenny (2009) emphasizes four such virtues in specifying the appropriate perspective that humans should take toward nature—reverence, humility, gratitude, and awe. The attitudes or virtues of reverence and humility flow from an acknowledgment that nature sets "a boundary or limit to human activity beyond which the latter exceeds the scope God has given it." Humility flows from the recognition that nature "stands over our intentional activity" and "confronts us with the limitations of our activity." Gratitude is an appropriate response to the awareness that nature, "precisely as that which is not the product of human intentional activity, is a gift that should be received gratefully" (p. 158). The fourth disposition is awe or wonder—appropriate reminders that "not everything in the world is to be made conformable to our will, use, or desire" (McKenny 2009, 159).

The "playing God" metaphor shares affinities with some recent prominent secular critiques of biotechnology. For example, Michael Sandel and Leon Kass link the acceptance of bodily limits to the inculcation and development of certain dispositions and virtues. Sandel criticizes efforts to move beyond therapy to the technological enhancement of human traits as a failure to accept the "gifted" character of naturally occurring human powers and achievements (Sandel 2004). Kass has argued in numerous writings that both the capacities and the constraints of our natural embodiment, especially the fact of human finitude, provide the most appropriate context for authentic human flourishing (Kass 2002). From a somewhat different vantage, Francis Fukuyama raises concerns about the ways that biotechnological alterations of human traits may challenge links between the shared biological basis of human nature and considerations of human rights and dignity. Because Fukuyama views human nature as the appropriate basis of shared human rights and dignity, he worries that some forms of biotechnological alteration may, over time, threaten our sense of a common humanity (Fukuyama 2002). Sandel, Kass, and Fukuyama each raise issues resembling those that arise in discussions of nature from more overtly religious perspectives. Sandel's emphasis on natural giftedness has affinities with religious perspectives that enjoin attitudes of reverence, care, and stewardship. Kass's interest in acknowledging the value of biological finitude corresponds, to a significant extent, with expressly religious concerns that recommend humility, caution, and the avoidance of hubris. Fukuyama's concern with equal rights and dignity resonates with religious perspectives that emphasize the requirements of stewardship and justice.[5]

The key point for my purposes is not whether Sandel's, Kass's, and Fukuyama's arguments stand or fall, but that their arguments are connected to and rooted in worldviews that themselves have deep and long cultural roots. But those connections do not settle things. Given the complex interplay between concepts of nature and other moral/religious values that may, in any particular case, provide the primary basis for particular judgments, it is in fact difficult to know exactly what to make of the overall thrust of the "playing God" metaphor, in either its religious or its secular versions. The same complexities of definition and interpretation noted in the first part of this essay beset efforts to specify precisely where and how nature establishes a border beyond which the language of transgression is unambiguously applicable. Overtly religious critics of Promethean actions or attitudes can be countered by religious scholars who interpret human alterations of or additions to nature as appropriate extensions of human responsibilities as created co-creators with God. Moreover, Sandel and Kass, in particular, have been challenged on two key matters: first, their reliance on the problematic distinction between therapy and enhancement; second, their shared tendency to pose the issues raised by biotechnology in a rather binary fashion: nature as given or nature as mastered; nature as gift or nature as object of manipulation; children as progeny or children as products, and so forth. Critics argue that these broad dichotomies are insufficiently nuanced to illuminate the full range of values and attitudes we bring to bear in assessing any *particular* development in biotechnology.

In all of this, one is struck by the persistent ambiguity that plagues efforts to invoke nature and the natural to do ostensibly normative work, and by the need in any particular case to buttress the appeal to nature with bridging arguments that are themselves open to dispute, often on independent grounds. Granted, it is important to acknowledge the range of efforts in ethics and the humanities, in philosophy and religious studies, to explore the relevance of concepts of nature for normative judgment. In much of that discussion, however, what seems most obvious are the differences drawn between discussions of "human nature" and "nature in general." Given such differences, and in light of the complexity of concepts of nature and the natural, it would be both desirable and useful to articulate some general concept of what it means to be human that could guide or constrain ethical choices in biotechnology. However, it is far less clear, from the vantage of either philosophy or religious studies, that we are warranted in appealing to "nature" in a way that provides clear ethical guidelines for current and proposed interventions, alterations, or additions to it.

Intrinsic Moral Concerns and Public Policy Choices

The deep roots of appeals to nature give them considerable moral heft. At the same time, the complexities in deploying the concept, and the tensions among the many different lines of thought about nature, must also raise concerns about the relevance of appeals to nature for policy deliberations about synthetic biology. It is hard to see how appeals to nature can function univocally to provide a basis for either a general ban, or even significant restrictions on, developments in synthetic biology as a discrete area of policy concern. Policy judgments on specific projects, including their public funding, will be most appropriately made on a case by case (or category by category) basis; such judgments will, of necessity, rely primarily on the more accessible and quantifiable basis of physical risks and benefits associated with specific proposals.

Several observations about appeals to nature in the debate about synthetic biology can be offered at least tentatively. First, appeals to nature and the natural capture a range of concerns that function less as dispositive judgments than as the intuitive expressions of broader worldviews.

I have suggested that the wide variety of definitions at work in such appeals, both descriptively and prescriptively, makes it hard to distill a set of broadly shared principles about appropriate policy on synthetic biology. Moreover, because appeals to nature generally function at the level of general sensibility, rather than as clearly articulated premises within carefully reasoned arguments, their function in the crafting of pluralistic policy is likely to emerge in tension with the requirements of public reason and liberal neutrality.

Second, appeals to nature and the natural are often linked to particular religious traditions that diverge widely in their respective interpretations of the conceptual links (or lack thereof) between the divine and the human; the links (or lack thereof) between facts and values; and different judgments about the appropriate range of human responsibility and stewardship, ranging from fairly conservative perspectives to more liberal interventionist stances. In light of that diversity, establishing a working consensus among traditions will prove difficult.

Third, while some critics target synthetic biology as different in kind from earlier forms of biotechnology involving genetic engineering (Boldt and Müller 2008), the distinction appears, at best, to be conceptually murky. There are two difficulties with the critique of synthetic biology as an illicit conflation of nature and artifice. As is obvious from studies in population biology, interactions between natural givens and human cultural and technological activities have occurred since the time of organized agriculture, if not before, making it difficult to describe biological givens in ways that do not incorporate such interactive effects. Moreover, insofar as the organisms of synthetic biology *add to*, rather than alter, neo-Darwinian nature, the bare appeal to nature, invoked conservatively, may be less applicable to synthetic biology than to other forms of biotechnology.

Fourth, it is useful to consider whether synthetic biology is more similar than dissimilar to other forms of cultural shaping and control that are not deemed controversial. Human culture is itself a natural expression of human agency, and the increasingly technological and scientific character of human culture does not appear to pose, in itself, distinct questions either conceptually or normatively. We are made ever more aware of the ways that cultural activity already shapes and reshapes human biology. Literacy, for example, has measurable effects on brain development and structure; the brains of piano virtuosos and London taxi drivers are materially altered by practices that we choose to situate on one side of the nature-culture divide. Thus, while there may be reasons for concern about launching full-scale efforts at genetic or cognitive enhancement, the fact that culture and nature already interact in ways that influence each other problematizes efforts to keep them hermetically sealed as conceptual categories.

Fifth, the polyvalency of concepts of nature suggests that arguments which seek to deploy them univocally will be made for purposes that are more rhetorical than analytic. To say that nature has many descriptive meanings, with varying prescriptive implications, is to acknowledge the obvious truths of the matter. Nature is *both* a source of the sublime and a garden to be tilled. Nature is *both* "red in tooth and claw" and an inspiration for Transcendentalist rhapsodies. Nature is, for Calvin, *both* the "theater of God's glory" and a realm disordered by the fall of Adam. Nature is *both* a wilderness to be respected and protected and a source of resources for human benefit. The pluralism of such conceptual groupings could go on for pages. Yet with each seeming dichotomy, we reaffirm both the fecundity and the elusiveness of nature itself as a category.

What, then, finally, of efforts to identify *the* meaning of nature for purposes of moral judgment or policy choice? Here, candor is called for. We should not expect so complex a term to serve as some lowest common denominator; efforts simply to stipulate its meaning will inevitably confuse as much as clarify. A better approach will be to acknowledge that the human relationship to nature reflects a range of values that we associate with the term. Indigenous peoples looked at nature in deeply animistic terms, even as they fashioned tools and weapons. Organic farmers respect natural patterns even as they disrupt them. We appreciate the "mysteries" of the genome even as we seek ways to regulate genetic science through patents and cooperative research and development agreements. Our alterations of or additions to the natural world need not be viewed in zero-sum fashion. "Nature" is not diminished by human creativity; at least it need not be.

Why, then, should synthetic biology give us particular pause? Perhaps, all things considered, it should not. But we are creatures of both reason and emotion, of worldview and of mastery, of art and technology. Nature is both constitutive of, and apart from, ourselves. Perhaps synthetic biology is merely the latest reminder of the unsettling tasks that face us as postmoderns. Prospective developments in biotechnology and synthetic biology confront us with the difficulties of invoking classical dichotomies to grapple with current topics. As John Hedley Brooke observes, "We are currently living through an era in which the traditional dualities [of nature/man, nature/supernature, natural/unnatural] are failing and our grasp of what it means to speak of 'nature' at all is wavering" (Brooke 2009). Our very "forms of life," as Wittgenstein might put the matter, are being transformed, and the choices are increasingly ours.

Notes

1. Quoted in Ronald Cole-Turner, "Synthetic Biology: Theological Questions about Biological Engineering," in *Without Nature? A New Condition for Theology*, ed. David Albertson and Cabell King (New York: Fordham University Press, 2010), 136–151, citation at 137.

2. The Ford Foundation project, which involved five years of research by more than 50 scholars, eventuated in the publication of two volumes edited by B. Andrew Lustig, Baruch Brody, and Gerald P. McKenny: *Altering Nature,* vol. 1, *Concepts of "Nature" and "The Natural" in Biotechnology Debates*; and *Altering Nature,* vol. 2, *Religion, Biotechnology, and Public Policy* (Dordrecht: Springer, 2010).

3. See, e.g., Londa Schiebinger, *Nature's Body: Gender in the Making of Modern Science* (Boston: Beacon Press, 1993).

4. For an especially cogent recent analysis of the ways that nature is invoked in biotechnology discussions, see Gerald McKenny, "Nature as Given, Nature as Guide, Nature as Natural Kinds: Return to Nature in the Ethics of Human Biotechnology," in Albertson and King, *Without Nature?*, 152–177.

5. See, e.g., Lisa Cahill, "Nature, Change, and Justice," in Albertson and King, *Without Nature?*, 282–303.

References

Agar, Nicholas, David M. Lodge, Gerald McKenny, and LaReesa Wolfenbarger. 2008. Biodiversity and biotechnology. In *Altering Nature,* vol. 2, *Religion, Biotechnology, and Public Policy*, ed. B. Andrew Lustig, Baruch A. Brody, and Gerald McKenny, 285–321. Dordrecht: Springer.

Andrianantoandro, E., S. Basu, D. K. Karig, and R. Weiss. 2006. Synthetic biology: New engineering rules for an emerging discipline. *Molecular Systems Biology* 2 (1):E1–E14.

Benner, Steven A., and A. Michael Sismour. 2005. Synthetic biology. *Nature Reviews Genetics* 6 (7):533–543.

Boldt, Joachim, and Oliver Müller. 2008. Newtons of the leaves of grass. *Nature Biotechnology* 26:337–339.

Brooke, John Hedley. 2009. Should the word "nature" be eliminated? In *Envisioning Nature, Science, and Religion*, ed. James Proctor, 312–336. West Conshohocken, PA: Templeton Press.

Dabrock, Peter. 2009. Playing God? Synthetic biology as a theological and ethical challenge. *Systems and Synthetic Biology* 3:47.

Fukuyama, Francis. 2002. *Our Posthuman Future: Consequences of the Biotechnology Revolution*. New York: Farrar, Straus, and Giroux.

Gaard, Greta (ed.). 1993. *Ecofeminism: Women, animals, and nature*. Philadelphia: Temple University Press.

Henderson, Martha. 2009. Re-reading the landscape: Death, redemption and resurrection on a Greek island. In *Envisioning Nature, Science, and Religion*, ed. James Proctor, 205–226. West Conshohocken, PA: Templeton Press.

Hume, David. 1973. *A Treatise on Human Nature*. Oxford: Clarendon Press.

James, William. 1982. *The Varieties of Religious Experience*. New York: Penguin.

Kass, Leon. 2002. *Life, Liberty and the Defense of Dignity: The Challenge for Bioethics*. San Francisco: Encounter.

Livingstone, David. 2009. Locating new visions. In *Envisioning Nature, Science, and Religion*, ed. James Proctor, 103–127. West Conshohocken, PA: Templeton Press.

Lovin, Robin, and Frank Reynolds. 1985. *Cosmogony and Ethical Order*. Chicago: University of Chicago Press.

Lustig, Andrew. 2009. The vision of malleable nature: A complex conversation. In *Envisioning Nature, Science, and Religion*, ed. James Proctor, 227–244. West Conshohocken, PA: Templeton Press.

McKenny, Gerald. 2009. Nature as given, nature as guide, nature as natural kinds: Return to nature in the ethics of human biotechnology. In *Without Nature? A New Condition for Theology*, ed. David Albertson and Cabell King, 152–177. New York: Fordham University Press.

Merchant, Carolyn. 1980. *The Death of Nature*. San Francisco: Harper and Row.

Merchant, Carolyn. 1995. *Earthcare: Women and the Environment*. New York: Routledge.

Midgley, Mary. 2003. Criticizing the cosmos. In *Is Nature Ever Evil?*, ed. William B. Drees, 11–26. New York: Routledge.

Ortner, Sherry. 1972. Is female to male as nature is to culture? *Feminist Studies* 1 (2):5–31.

Peters, Ted. 1995. "Playing God" and germline intervention. *Journal of Medicine and Philosophy* 20 (4):365–386.

Proctor, James. 2009a. Introduction: Visions of nature, science, and religion. In *Envisioning Nature, Science, and Religion*, ed. James Proctor. West Conshohocken, PA: Templeton Press.

Proctor, James. 2009b. Environment after nature: Time for a new vision. In *Envisioning Nature, Science, and Religion*, ed. James Proctor, 293–311. West Conshohocken, PA: Templeton Press.

Rolston, Holmes. 1988. *Environmental Ethics: Duties to and Values in the Natural World*. Philadelphia: Temple University Press.

Ruether, Rosemary Radford. 1992. *Gaia and God: An Ecofeminist Theology of Earth Healing*. San Francisco: Harper and Row.

Sandel, Michael J. 2004. The case against perfection: What's wrong with designer children, bionic athletes, and genetic engineering. *Atlantic Monthly*, April, 51–62.

Schiebinger, Londa. 1993. *Nature's Body: Gender in the Making of Modern Science*. Boston: Beacon Press.

Tuomela, Raimo. 1995. Feminist philosophy. In *The Cambridge Dictionary of Philosophy*, ed. Robert Audi, 263. Cambridge: Cambridge University Press.

Zoloth, Laurie. 2008. Go and tend the earth: A Jewish view on an enhanced world. *Journal of Law, Medicine, and Ethics* 36 (1):10–25.

2

Creating Life: Synthetic Biology and Ethics

Joachim Boldt

The invention and development of recombinant DNA techniques in the 1970s led to what is nowadays known as "genetic engineering" and to "genetically modified organisms" such as transgenic maize, insulin-producing bacteria, and the oncomouse. Synthetic biology is usually thought of as a more recent development. It may come as a surprise, then, that the term "synthetic biology" entered the scientific scene around the same time as "genetic engineering."

In 1978, the molecular medical scientists Waclaw Szybalski and Anna M. Skalka wrote in an editorial of the journal *Gene*, "The work on restriction nucleases not only permits us easily to construct recombinant DNA molecules and to analyze individual genes but also has led us into the new era of 'synthetic biology' where not only existing genes are described and analyzed but also new gene arrangements can be constructed and evaluated" (Szybalski and Skalka 1978). Szybalski and Skalka thus identify the advent of recombinant DNA techniques with a substantial turn in the development of biology. Whereas biology before the invention of recombinant DNA techniques had been an analytic and descriptive science, they claim, it became afterward a science capable of constructing new sets of DNA.

This way of conceptualizing scientific progress in biology is analogous to how progress in chemistry often is described. In the early twentieth century, when reliable and reproducible chemical processes for producing complex molecules from simpler ones were developed, these techniques were labeled "synthetic chemistry" to distinguish them from a precursory, purely descriptive, and analytic period of chemical research. In this vein, the biochemist and former secretary general of the European Research Council, Ernst-Ludwig Winnacker, remarked in a presentation in 1986, "A development looms at the horizon that one may term, analogous to the course of events in chemistry, 'synthetic biology.' It promises to initiate a similar impact on research and development" (Winnacker 1990, 371; my translation).

Viewing the shift from analysis to synthesis as an expression of scientific progress highlights the fact that knowledge generated in the natural sciences is instrumental knowledge. Knowing that a certain process will under given conditions lead to certain results allows one to steer these processes in a targeted and controlled manner. Thus, knowledge attained in the natural sciences is a tool for manipulating objects and events. What is more, synthesis—that is, the manipulation of events—is not only a by-product of correct analysis, it is itself the measure of whether or not analysis is correct.

Besides the shift from analysis to synthesis, comparing advances in biology to earlier progress in chemistry also expresses the idea that the natural sciences are ordered hierarchically from the bottom up. The laws of the physical world are meant to form the basis of the laws of chemistry and biology. In turn, one expects to be able to explain biological events such as intra- and intercellular processes fully in terms of chemical and physical concepts.

From this early perspective, the synthetic biology to come is characterized by two main features: First, it represents the shift from mere observational analysis to the ability to build, initiate, and steer DNA-based processes. Second, it carries scientific progress forward from the realm of chemistry to the realm of biology. Synthetic biology, it is supposed, will lead us into an era of life as an object of industrial production. As a consequence, synthetic biology is thought to be not merely one of many new fields of research, but the spearhead of the general course of scientific progress.

Synthetic Biology Today: Creation and Reduction

These announcements notwithstanding, it took about two decades before synthetic biology actually was born and baptized and gathered pace. The key factor in this development was the increasing power of gene sequencing and synthesis technologies (Carlson 2003). While genetic engineering was restricted to transplanting short strains of DNA into a host organism, thereby adding or changing single features of an otherwise stable, existing organism, the new ability to synthesize genomes at full length opened up the possibility to alter the whole DNA-related behavior of prokaryotic single-cell organisms at once (J. Craig Venter Institute, n.d.).

At first sight, this may amount to a purely quantitative difference, without further significance—a numeric expansion from genetic engineering's plasmids of around 5,000 base pairs to synthetic biology's strings of 600,000 base pairs. Nonetheless, quantitative changes do often coincide with qualitative differences. The classic example is the sorites paradox, according to which adding a grain of wheat to an existing number of grains never can be said to turn the existing number of grains into a heap of grain. Still, one would not want to conclude that it is impossible to build a heap by adding grain to grain. Obviously, the semantics of "heap" does not depend on the exact number of grains involved, but on the way in which this kind of object enters into our world of experiences. A heap is a visible elevation, one can stumble over it, and so on. In a parallel way, the quantitative increase in DNA synthesis may very well make a difference with regard to how well one is acquainted with the resulting objects or, generally speaking, with regard to how smoothly one is able to integrate the ability to synthesize DNA into one's common understanding of the effects of one's actions. This holds true even if no clear-cut line in terms of the number of base pairs can be drawn between "single grain" genetic engineering and "heap" synthetic biology.

Synthetic biology's heap is its appeal to creation. Portrayals of synthetic biology research are full of phrases such as "writing the software of life," "assembling living organisms," "building life," and "Nature 2.0," none of which can be found in characterizations of genetic engineering. In comparison to genetic engineering, synthetic biology constitutes a shift from the paradigm of manipulating existing organisms to creating novel entities.

At this point, though, one has to proceed carefully. If the phrase "creation of life" is used without qualification, it does not square with most current research in synthetic biology, which by and large follows a "top-down" strategy in that it uses the parts and pieces of already existing organisms. One strand, for example, is aimed at eliminating genes from bacteria with very small genomes, leading eventually to a so-called minimal genome that contains only the genes necessary to keep the organism functioning. A minimal genome is meant to be a platform organism that can be equipped with tailor-made synthetic DNA, and it would be a significant accomplishment, but cannot be described as the creation of life; it presupposes the existence of life. There are also "bottom-up" approaches that are aimed at building artificial cells completely from single molecules, but these approaches are considerably further from realization than the top-down approaches, and even if they were eventually successful, it could still be debated whether life had been created in the same way in which one can ascertain that a car or any other kind of artifact has been created. Life, it may seem, is a kind of property that cannot be directly produced, but emerges under certain conditions that one may identify and bring together.

With regard to a specific organism, though—rather than to the general phenomenon of life—speaking of an act of creation can be meaningful, even in the case of top-down research. Transplanting a synthetic genome of a known organism will not suffice, but suppose, at the other end of the scale, that a genome has been devised that has no natural template and exhibits features that natural organisms are not capable of developing. In this case, speaking of a novel organism or a novel form of life becomes perfectly sensible, the many ambiguous cases in between these two end-points of the scale notwithstanding.

Events such as the International Genetically Engineered Machine competition confirms this view. The iGEM competition convenes groups of undergraduate students from industrialized countries to engineer simple bacterial organisms that carry out specific tasks (iGEM Foundation, n.d.). In order to do this, iGEM teams are asked to make use of and contribute to an open source registry of standardized biological parts, the so-called BioBrick parts. Having created blueprints of the relevant DNA sequences, the teams then synthesize and transplant the DNA strands into suitable bacteria. Although iGEM organisms up to now must count as modified—as opposed to novel—bacteria, the competition's take on the modification task is unique, since the point of departure is not an identified shortcoming of a given organism but a set of "bricks" ready to be combined in any way one may judge worthwhile. The (intentional) similarity of the name BioBrick to "Lego brick" is by no means superficial. Just as building something out of Lego bricks is an ingenious, creative task, building DNA sequences out of BioBricks is an act of inventive engineering. Even if the output may not qualify as a novel organism, the perspective is set: It is the perspective of rational, technological creation.

The perspective of creation in today's synthetic biology resumes the early expectations concerning a shift from analysis to synthesis. Undoubtedly, genetic engineering also involved synthesizing DNA sequences and implanting them into host organisms. It was a very limited ability, however. A meaningful ability to synthesize DNA—an ability worthy of being called "synthesis"—is achieved only in synthetic biology. Think of the common metaphor of a text: whereas genetic engineering resembles editing a text, synthetic biology sets in with a blank sheet of paper. The transition from descriptive biology to genetic engineering to synthetic biology can thus be paraphrased as the development from reading to editing to writing DNA.

The title of the iGEM competition highlights a second characteristic of synthetic biology. For an outside observer, it may be puzzling to learn that a "genetically engineered machine" is identical to what he or she otherwise thinks of as a bacterial organism, albeit a genetically modified one. The expression "genetically engineered machine" conflates the realm of mechanics and the realm of living beings, thereby obscuring one of the most central, everyday life distinctions of human discourse. "Living machine" is another example in the same vein of how organisms containing synthetic DNA tend to be understood in synthetic biology (Bleicher 2010). One source of these apparently self-contradictory phrases is the importance attached in synthetic biology to principles of engineering. On the one hand, engineering typically results in machinelike artifacts; on the other hand, synthetic organisms are undoubtedly living organisms; hence the products of synthetic biology are both machinelike and alive. Another source lies further back, though. Since the emphasis on engineering is itself a symptom of the expectation that synthetic biology is going to repeat the trajectory of chemistry, which started with analysis and advanced to chemical engineering, expressions such as "genetically engineered machine" reflect the goal of reducing the basic processes of life to chemical processes. Just as chemistry showed it had succeeded at analyzing substances because it went on to succeed at synthesizing them, so synthetic biology aims to give a final analysis of life—to show that biological organisms can be explained by reference to their constitutive parts—by demonstrating that they can be synthesized.

Reductionism is not always worrisome. Quite the contrary: understood as a methodological hypothesis, reductionism can be a fruitful starting point for generating research questions and attaining research results that suit one's practical purposes. Nonetheless, if reductionism is taken to represent the only way in which life can possibly be described, it can interfere with alternative accounts of how life ought to be understood and dealt with. In other words, if reductionism turns ontological, it needs to be evaluated from a broader, not solely scientific perspective.

Attitudes, Acts, and Ethics

In his writings on biotechnology in the 1980s, the philosopher and bioethicist Hans Jonas anticipated the prospect of living machines. In biogenetic technology, he wrote, "the work of the hands of homo faber . . . literally gains life of its own and a certain degree of autonomy. At this threshold—the potential point of origin of widespread growth and diffusion—it does indeed befit him to pause and reflect on the basic precepts of his doing" (Jonas 1987, 205ff; my translation).

Adapting Jonas's admonition to synthetic biology, one is first of all reminded of the textbook distinction between theories of ethics that are oriented toward acts and consequences and theories that focus on good-will, attitudes, and virtues. In current discussions on ethical aspects of synthetic biology, clearly the focus on acts and impacts dominates. Bio-safety and biosecurity issues, the two main topics of the prevailing debate, both belong to this category of ethical reflection.

Jonas, however, urges us to turn our attention in a different direction. In reflecting on the basic precepts of one's doing ("zu grundsätzlicher Besinnung innezuhalten," as the German text has it), one is first of all concerned with the attitudes and grounding assumptions that guide action. It is this perspective that will be pursued first in this chapter, since only from this perspective can we obtain a broad understanding of what the actions to be examined actually consist of. Subsequently, an attempt will be made to bridge the gap from the analysis of basic action-guiding assumptions to issues of risk assessment.

In the case of synthetic biology we must ask, What is the ethical significance of conceiving of oneself as a creator of life rather than as a manipulator of existing organisms? For the moment, let us postpone this issue and turn our attention to the second characteristic of synthetic biology: its reductionist analysis of the phenomenon of life. This analysis is itself part of a particular attitude toward reality in general. In contrast to relations between persons, the ideal of which is mutual understanding and sharing of reasons and values, the ideal that comes to the fore in synthetic biology is an observational stance, the stance of a neutral investigator who examines the phenomenon of life unaffected by any distorting assumptions. In continental philosophy, the distinction between an observational attitude and an attitude of shared perspectives gets to the heart of the difference between positivism and hermeneutics. In analytic philosophy, the same issue has surfaced in debates about first-person versus third-person perspectives.

The ethical implications of this dichotomy are especially striking when the entities under observation are persons. Nonetheless, there are also important ethical implications when life in general is the object of observational scrutiny. Since synthetic biology is the heir of the observational, physico-scientific ideal of science in the realm of life, it is worth taking a closer look at the ethical significance of this stance as a first step in mapping the ethical terrain to which synthetic biology belongs.

Observing Life

In our everyday encounters with one another, we follow an assortment of ontological assumptions about what we are capable of. These assumptions are seldom explicit, but they are of utmost ethical importance. First of all, we assume that we are capable of having reasons for our actions. Reasons tell us why we suppose certain actions to be good. In other words, getting to know these reasons helps us both to comprehend why someone acts as she or he does and to reflect upon our own stance toward these reasons and actions. What is more, the ability to act for reasons presupposes the freedom to choose otherwise. So-called compatibilist theories of free will notwithstanding, the inherent connection between acting for reasons and freedom and their joined opposition to deterministic explanations remain a matter of fact of our human self-understanding.

These assumptions are ethically relevant because they establish humans as persons who are able to act responsibly. It would simply not make sense to try to ethically comprehend and evaluate the acts of someone whose behavior does not result from freely chosen reasons. In addition, when we take for granted the existence of reasons, we are drawn into seeking a shared perspective on what might be thought of as good in a given situation. Everyone who takes a stance on what is good is in principle considered equally capable of providing answers, and so everyone ought to be taken account of. Thus, conferring the ability to act for reasons is tantamount to assigning equal status to all those whom we deem capable of taking this stance. In a nutshell, freedom and coexistence are the hallmarks of the framework of mutual understanding.

In contrast, in the framework of observational explanation, one deploys the concepts of cause and effect in order to account for specific forms of behavior and movement. Cause and effect are thought of as pure matters of fact, which is why they do not stand in any relation to attempts to attain the good, at least as long as attaining the good is understood as an epistemic task. In addition, cause and effect are supposed to provide deterministic understandings. As a result, the relation of the explaining scientist to the entities whose behavior is explained is one-sided. While the scientific observer can think of himself as using the explanations gathered to inform his preferences and thus perform ever more sophisticated and efficient actions, the entities explained appear to be completely subject to powers that are insurmountable for the entities themselves. Thus, necessity and control are the two notions central to the framework of observational explanation.

In encounters between persons, these two starkly opposed perspectives are plainly of great ethical significance, since reinterpreting *reasons* as *causes* in effect removes the status of *fellow person* from the one whose behavior is under consideration. This is the core ethical challenge of medical research on humans. It is also the reason why attempts to explain deviant behavior purely medically—in terms of neurobiological processes, for example—can come to seem inadequate.

It is less obvious, though, that this distinction also is relevant when the entities to be explained are not humans but nonhuman animals, let alone bacteria. In arguing for this, it is important to note, first, that the decision about which of the two frameworks ought to be employed for a given range of entities cannot ultimately be settled on the grounds of empirical evidence alone. For example, even if one does not grant primates the ability to communicate and reason consciously to the same extent as humans, one may still want to concede to them a basic form of an ability to gain insights. This judgment will be guided by empirical evidence, but empirical evidence alone will not convince someone who sticks to the framework of deterministic causal explanations that his account *must* be wrong. In the same vein, it is always theoretically possible to deny a person the ability to reason appropriately, and accordingly to explain actions with recourse to nonepistemic, "blind" forces, no matter how much she or he may protest. Concerning the question of status, empirical evidence serves as a heuristic, not as a proof.

What is more, we usually consider the ability to suffer pain a decisive criterion for bestowing moral status on the organism in question, and with good reason. Nonetheless, if one wonders why to choose just this criterion as a reason for moral obligation, answers such as "experiencing pain is not in the interest of the organism" will not suffice unless one presupposes that the suffering of the organism ought to matter to those who are perceiving it. In other words, the ability to suffer can become a reason for accepting the moral status of an organism, and obligations toward that organism, only if one at the same time allows the organism to belong at least partially to the class of subjects that are capable of sharing a perspective on what may count as enjoyable and good—a class that includes oneself. Thus, accepting the ability to suffer pain as a criterion for moral status presupposes the willingness to partially adopt the framework of mutual understanding.

We can dig further. Not only does accepting the ability to suffer pain as ethically relevant depend on adopting the framework of understanding, but detecting this ability itself presupposes a nonobservational point of view. The ability to suffer pain becomes visible only if one imposes one's own experiences of specific situations and reactions on the organism that one observes. The ontological assumption that someone suffers pain thus is rooted in the perspective of shared subjective and normatively relevant experiences.

There are further ontological attributes of organisms, in addition to the ability to feel pain, that are also widely held to originate in the framework of understanding. Ultimately, the most basic ontological characteristics introduced by this framework are *having an identity independent of one's physical makeup* and *being an organism that is distinct from its environment.*

Organisms constantly exchange their physical substrate with their environment without thereby losing or changing their organic identity. An organism's identity, therefore, does not hinge on the identity through time of its physical constituents, but depends on a process of continual self-renewal. In his philosophy of biology, Hans Jonas describes this phenomenon as the priority of form over matter in the realm of life (Jonas 1966, 79–83). Hence, ascribing organic identity to an entity is not a purely empirical statement but makes recourse to the experience of being such an organism oneself.

At first sight, it may appear to be no huge challenge to determine the line between an organism's body and its environment. In doing so, however, one must assume the existence of an antecedent, nonphysical identity of the organism. Without this presupposition, the apparent boundary turns out to be a mere surface phenomenon of what is in actual fact a united piece of reality that constitutes and determines both would-be organism and environment. Again, the assumption of the existence of an organism that is acting in an environment traces back to the experience of being this kind of entity.

Hans Jonas, who in parallel to the distinction between the scientific stance and the perspective of understanding contrasts the perspective of "the mathematician God" with the "testimony to the contrary" of the living body, relates the characteristics of having a nonphysical identity and having an environment to the notion of freedom: "And in this polarity of self and world, of internal and external, complementing that of form and matter, the basic situation of freedom with all its daring and distress is potentially complete" (Jonas 1966, 83). In this way, reason and freedom of the will, the ontological *differentiae specificae* of persons, can be conceived of as the full realization of tendencies that also govern the behavior of lower organisms. Reason, interest, and instinct form a descending line of behavior-guiding powers that become more and more rigid toward the bottom, but nonetheless all partake in constituting organisms that display self-renewal in exchanging parts with an independent environment.

If one is to look for the ethical core of the molecular life sciences, here it is. With regard to the phenomenon of life itself—life as a general phenomenon—taking up the observational stance amounts to using a telescope in order to get a better view of an object right in front of one. It becomes impossible to discriminate that which makes an organism an organism, and, if one introduces these ontological characteristics anyway, still the normative implications of these characteristics evade one's attention. Michael Hauskeller concludes, "The process of reification is never complete and remains largely conceptual and perceptual. Biotechnology just gives us the means to consolidate our blindness towards the independent reality of an animal's existence" (Hauskeller 2007, 100).

Synthetic Biology and the Normative Content of Life

A reductionist approach to life may not be worrisome when it is only a methodological hypothesis, but if it is taken to be an ontological rather than a methodological statement, it cuts off the normative aspects of the concept of life. While this will not disturb many people as long as synthetic biology is directed toward single-cell organisms, it is bound to become an issue of greater concern when synthetic biology is applied to higher animals or human cells. To be sure, the synthesis of a human being will not happen in the foreseeable future. The man-made homunculus is a literary motif, not the next technological invention. Instead, the approach to higher organisms will happen more incrementally and slowly, as synthetic biology's methods of rational design, standardization, and modularization are applied to somatic cells. Even so, we will have to ask ourselves how we can strengthen the culturally embedded understandings of the normative aspects of life against the reductivist aspirations of synthetic biology backed up by the authority of scientific truth. Bearing in mind that one is not forced to switch to the perspective of understanding by brute facts, regardless of what kind of organism one is going to analyze, the challenge amounts to deciding at what step on the ladder of life one is going to insist on the inherent normative value of the entity under scrutiny.

Ontology, Creation, and Risk Assessment

If life is always, albeit to differing degrees, in possession of a capacity of self-renewal, exchange, and freedom, and if these attributes can be identified only through a perspective of understanding, then synthetic biology will never fully understand life. In particular, synthetic biology is apt to lose sight of the unpredictability of the alterations and struggles that organisms are exposed to in the course of their development as individuals and as members of a species.

Practicing the subtle art of growing microbial culture can serve as a constant reminder of this kind of unpredictability—an unpredictability, by the way, that one usually copes with by nourishing and cherishing the culture. This kind of reaction is a further hint that encountering life even in its most basic forms is, indeed, an encounter with what Jonas called "daring and distressing" freedom. From the interpretive perspective of synthetic biology, of course, such reactions appear to be based on prescientific illusions. The more synthetic biology is immersed in engineering methods and ideals, the more it will cling to this line of thought.

Besides quarrels in the lab, dealing with unpredictability is of great importance in assessing the risks, to human health and the environment, of uncontained uses of synthetic and genetically modified organisms. For this reason, the debate about risks, usually held in consequentialist terms, is intimately connected to broader philosophical and ontological issues. Conflating organisms and machines can constrain one's judgment as to what kinds of effects a synthetic organism may have on its environment. At the same time, the engineering perspective can also lead to underestimating the animate environment's resilience—its ability to integrate foreign forms of life and recover from harm. Still, justifying the intentional release of synthetic organisms, the effects of which on the environment are unknown, by reference to nature's integrative abilities is different from deciding against release on grounds that it might lead to uncontrollable spreading. Even if one grants nature's ability to integrate novel organisms, releasing novel organisms still stands in need of strong justificatory reasons since the interactions of a novel organism with its environment remain unpredictable and carry risks. It is wise not to run these risks unless one is compelled to do so by overwhelming benefits.

Synthetic biology's creative attitude reinforces the relevance of these concerns about risk taking. The creative and inventive aspects of synthetic biology are reminiscent of artistic activity. As in art, the free play of the imagination can be part of the production process, the resulting object need not serve any immediate purposes, and given the easy availability and moderate size of the necessary equipment, synthesizing organisms can be accomplished by small groups, even by individuals. If, in addition, the synthetic objects were directed at a human audience for reflection and edification, not much would be left to distinguish synthetic biology from proper art. Indeed, emerging labels such as "BioArt" and "Synthetic Aesthetics" bear witness to this parallel. Given that the products of synthetic biology are alive and usually not meant to be displayed in museums, public anxiety about the new technology is understandable. The freedom of the artist who spares no thought on utility, combined with the artwork being a procreative organism, is a mixture of fragmentary perceptions bound to cause worry.

Obviously, this is a vision, not reality. Despite its potential kinship to the arts, synthetic biology research as it takes place in the lab today is conducted as a biotechnology, with carefully selected and beneficial applications. What this vision brings to the fore, though, is the fact that synthesizing novel organisms is an achievement that can fairly fittingly, albeit metaphorically, be described as "giving birth to" an entity, as opposed to "producing" or "assembling" it. The reason is that the aspects of novelty and aliveness evade the meaning of the two latter verbs while both find appropriate expression in the former phrase.

The logic of the idea that science is advancing toward creative synthesis encourages one to think that having reached the ability to synthesize life is tantamount to having climbed the peak of explaining and manipulating life. If one designs and assembles all the parts of a new product, one expects the product to function in the way it is devised for, and further expects to be able to give an account of why it does so and likewise—in case the products fails to meet the expectations—to be able to localize and explain the defect. But even if the bottom-up approaches in synthetic biology (the only agenda in synthetic biology that comes close to creating life entirely from nonliving basic parts) one day were to be successful, life may turn out to be a kind of property that does not allow of such epistemological transparency. Having turned into homo creator, the former homo faber may find his aspirations and expectations proven wrong by the phenomenon of life. This is to say, in effect, that his predictions about the future behavior of the novel organism may turn out to be less reliable than homo faber himself is prepared to expect.

Combining novelty and aliveness thus adds to the challenge of risk assessment. This holds for bottom-up as well as top-down approaches. If top-down synthetic biology ever assembles an organism that deserves to be called a "novel form of life"—like an artist creating a new, original piece of work—interpreting the risks that this organism may pose to its environment becomes especially complex. While risk assessment of uncontained uses of old-fashioned genetically engineered organisms can be based on existing knowledge about the behavior of the host organism receiving the gene sequence in question, considerably less of this kind of knowledge will be available in the case of novel organisms. As a consequence, generating reliable rules and methods for risk assessment will be one of the most pressing endeavors in order to promote the responsible utilization of synthetic organisms.

Contextualizing Synthetic Biology

Synthetic biology has its origin in a specific idea of scientific advancement that aims at reductive explanations of life and incorporates a self-understanding geared to the ideal of a technological homo creator. Regardless of whether one agrees with this diagnosis, one obvious path toward making policy recommendations about synthetic biology is to explore what these characteristics mean for the task of ensuring safe and beneficent applications. One may then argue about how cautious one ought to be in utilizing the novel objects—wary and more cautious, if one shares the diagnosis, confident and less cautious if one is not in agreement about this analysis of synthetic biology's roots. As a consequence, bringing together, on the one hand, the basic attitudes and grounding assumptions that guide actions and, on the other hand, the safety issues related to judging the consequences of actions can help to identify and make intelligible much of what propels the public debate about synthetic and genetically modified organisms.

As a complement, however, a second path is left to be examined: The specific understanding of life and of the biologist's activity characteristic of synthetic biology may shape and guide research and development. Thus, one may ask what synthetic biology looks like from a perspective in which "life" is more tightly connected to ideas of value and development. Some thoughts along these lines are meant to complete the ethical analysis provided here.

Let us suppose that the primary aim of gaining biological knowledge is to come to terms with the interactions of *subjects*, as opposed to explaining the internal causes of an individual *object's* behavior. If one adopts this perspective, one takes for granted the subject as a given condition of one's examinations and investigates the specific ways in which it reacts to and acts upon the environment, thereby getting to know its tendencies, inclinations, instincts, or reasons, depending on the subject's kind.

Making use of a subject for one's own purposes can, in this view, amount to recognizing lucky coincidences of the subject's abilities and of one's own interests, preparing situations in which these abilities manifest themselves, and finally convincing the subject to cooperate, if the subject is of a kind for which this is a meaningful possibility.

In the case of microorganisms, research into microbial ecosystems will be first on the list, not just in order to evaluate an organism's effect on the environment (as in the case of field release of synthetic organisms), but in order to better understand the organisms themselves, their development, and their actions and reactions toward one another and other parts of their environment. "Evolution is not an unfolding but an historically contingent wandering pathway through the space of possibilities," Richard Lewontin claims, and he concludes that evolutionary genetics ought to redirect its attention toward the coevolution of an organism and its DNA, on the one hand, and its environment, on the other hand—a process, he maintains, that cannot be dissolved into the deterministic processes of the unfolding of DNA (Lewontin 2000, 88, 125–129). This reorientation of synthetic biology from the internal causes of objects to the interaction of subjects can be understood as a case in point of Lewontin's claim.

If one is looking for ways to make use of microorganisms for one's own purposes, the first option that comes to mind is to systematically *search* for a suitable organism with beneficial traits. A second step will be to affect the development of a promising organism by changing its environment.

Proceeding from here to assembling synthetic organisms cannot be motivated by an urge to gain deeper insights into the nature of the organism, since gaining insight, on this account, is tantamount to getting acquainted with the organism in the course of its life. While synthetic biology's typical assumption is that the relevant knowledge concerning an organism and the possibilities of its behavior is contained in the ability to design and assemble it, the alternative account systematically envisages that we will continually be confronted with behavior of the organism that can be rendered plausible but not turned into fully predictable chains of cause and effect.

If one has important interests that no existing organisms—and no viable alternative technology—can meet, then synthesizing organisms can still be justified. Curing illness and securing an ecologically sustainable economy are two among the many valuable ends that researchers in synthetic biology frequently invoke to justify their work. They are right to do so, it seems, even if research is understood as aimed at mutual understanding rather than observation. However, adopting the perspective of mutual understanding reinforces the relevance of assessing soberly and accurately the benefits of synthetic biology applications in comparison with the benefits of alternative solutions to problems synthetic biology is meant to address. It remains an open question whether synthetic biology will prove to promote socially valuable ends better than alternative means. The future of synthetic biology is in finding positive answers to this challenge.[1]

Note

1. The German Federal Ministry of Education and Research supported this work as part of the national joint research project "Engineering Life: An Interdisciplinary Approach to the Ethics of Synthetic Biology" (01GP1003).

References

Bleicher, Ariel. 2010. How to build a living machine. *Scienceline*. http://www.scienceline.org/2010/01/how-to-build-a-living-machine/.

Carlson, Robert. 2003. The pace and proliferation of biological technologies. *Biosecurity and Bioterrorism: Biodefense Strategy, Practice, and Science* 1 (3):203–214.

Hauskeller, Michael. 2007. *Biotechnology and the Integrity of Life*. Aldershot, UK: Ashgate.

iGEM Foundation. N.d. Synthetic biology: Based on standard biological parts. http://ung.igem.org/About. Accessed September 12, 2012.

J. Craig Venter Institute. N.d. The first replicating cell: Overview. http://www.jcvi.org/cms/research/projects/first-self-replicating-synthetic-bacterial-cell/overview/. Accessed September 18, 2012.

Jonas, Hans. 1966. Is God a mathematician? The meaning of metabolism. In *The Phenomenon of Life: Toward a Philosophical Biology*, ed. Hans Jonas, 64–98. Chicago: University of Chicago Press.

Jonas, Hans. 1987. Mikroben, Gameten und Zygoten. Weiteres zur neuen Schöpferrolle des Menschen. In *Technik, Medizin und Ethik. Zur Praxis des Prinzips Verantwortung*, ed. Hans Jonas, 204–218. Frankfurt am Main: Suhrkamp.

Lewontin, Richard. 2000. *The Triple Helix: Gene, Organism, and Environment*. Cambridge, MA: Harvard University Press.

Szybalski, Waclaw, and Anna M. Skalka. 1978. Nobel prizes and restriction enzymes. *Gene* 4 (3):181–182.

Winnacker, Ernst-Ludwig. 1990. Synthetische biologie. In *Die zweite Schöpfung. Geist und Ungeist in der Biologie des 20. Jahrhunderts*, ed. Herbig Jost and Rainer Hohlfeld, 369–385. München: Hanser.

3

Engineered Microbes in Industry and Science: A New Human Relationship to Nature?

Gregory E. Kaebnick

With some scientific and technological developments, the public gets excited when the technology hits the streets and generates new products and services. Radio caught the public's attention when radios became available. The Internet had been in the works for some years before most people even knew about it. But with developments in biology, the excitement tends to precede the application. In the 1990s, genetic engineering was going to cure the incurable; fifteen years on, there are only a few scattered reports of success, and then only on a few individuals at a time and not completely smoothly (the treatment has on a few occasions led to cancer). In 2000, when the Human Genome Project succeeded in completing the first draft of a human genome, the results were going to change medicine; ten years later, the *New York Times* ran a story about how meager the results were (Wade 2010). With embryonic stem cells, we are now just past the first flush of excitement, and there's no reason to be disappointed that new cell-based treatments have not gone into clinical use yet. Still, it's plain that the excitement preceded the success (assuming it comes) by many years.

Anyone who has observed this history can be forgiven for adopting a wait-and-see attitude toward the branch of genetic engineering that has been dubbed "synthetic biology." With synthetic biology we are still caught up in the first flush of excitement. Nonetheless, there are reasons to think that some lines of work in synthetic biology might actually come to fruition and could have a far-reaching impact. First, synthetic biology has to date focused on simpler organisms, so that the biological changes it wants to effect are, though not simple, at least simpler. Second, because the organisms entirely lack sentience, there are no concerns about obtaining their informed consent to the research, nor even about harming them, and therefore some of the regulatory hurdles that slow down gene therapy, for example, are gone. In short, even if synthetic biology falls short of the most grandiose predictions—such as that it will lead to a second industrial revolution—it may well turn out to represent an important step forward in the human ability to design and build living organisms, and then put them to profitable use.

But this prospect raises other questions, including a threshold question about whether the very idea of designing, building, and using living organisms is troubling. The concern here is that synthetic biology fundamentally changes the human relationship to nature. This kind of concern has arisen repeatedly as biological science has advanced, but with synthetic biology it is arguably especially sharp. A fully adequate way of articulating this concern, however, has been as elusive in the context of synthetic biology as everywhere else. Often, the concern is expressed very bluntly and dramatically, as a question about whether it is appropriate for humans to take on the task of "creating life," whether by doing so they are "playing God," and this way of putting it has led to some heavy sighing and eye-rolling among the proponents of synthetic biology. Drew Endy, a leading figure in the field, has asserted in an interview that "the questions of playing God or not are so superficial and embarrassingly simple that they're not going to be useful" (Edge 2008). In the introduction to *What Is Life?*, a book tracking one particularly provocative line of work in synthetic biology, Ed Regis, a science writer and philosopher by training, agrees that the field raises moral, political, and legal questions, but mocks most of what has been said about them: "The problems started with the perennial and trite layman's taunt, the claim that creating life was 'playing God'" (Regis 2008, 6).

To synthetic biologists, the work is intrinsically good—it is a morally noble activity, in and of itself. The pursuit of science is the advancement of human mastery, and synthetic biology exemplifies human mastery especially dramatically: as Regis provocatively puts it, the creation of a brand-new form of life, as some lines of protocell creation aim to do, amounts to a "second creation." It is humanism at its most grandiose. The pursuit of science is also the pursuit of understanding, the refinement of human intelligence—a less self-aggrandizing but still grand endeavor. So understood, part of the very point of synthetic biology is philosophy in its original sense—love of knowledge that encompasses both physics and metaphysics. In fact, it's also philosophy in its current sense, which concerns the investigation of human systems of meaning. "The most profound and provoking question raised by the effort to build an artificial living cell," writes Regis, "was exactly the one that lurked as an unseen presence beneath all the rest: the age-old riddle, What is life?" (Regis 2008, 7).

I will argue here for a middle way on concerns about the very idea of designing, building, and using living organisms. We should take it seriously. This will mean exploring and examining the "playing God" concern—trying to articulate it more clearly. I will argue that language about playing God turns out to stand (in synthetic biology) for several different though related objections about how biotechnology may change the human relationship to nature. They are all interesting and potentially sophisticated positions, but they depend on different philosophical commitments, and they vary in the extent to which they are legitimate bases for public policy making. Some of them could, in some circumstances, legitimately have an effect on public policy, but none of them ultimately make a difference for public policy on synthetic biology as it stands now. If the field turns out to be as powerful as some envision, matters could change.

Metaphysical Mistakes: At Odds with the Cosmos?

One way of interpreting the thought that synthetic biology changes the human relationship to nature is to see it as making a fundamentally metaphysical claim. The language of "playing God," for example, suggests that we humans are in some way invading divine territory. In fact, there are three kinds of metaphysical objections that might be at work in "playing God."

First, one might object that synthetic biology simply shows certain metaphysical understandings of life to be wrong. In particular, one might contend that synthetic biology undermines the specialness of life by showing that life is a purely material phenomenon, that a living organism is just a complex combination of ingredients. This objection seems to be contemplated by advocates of synthetic biology more often than by critics. In the first scholarly article on the ethical issues of synthetic biology, for example, Mildred Cho and coauthors weighed the possibility that, by defining life in terms of DNA, synthetic biology reduces life to a single biological feature and therefore "may threaten the view that life is special" (Cho et al. 1999). It popped up again when scientists synthesized the genome of *M. mycoides*, inserted it into a *M. capricolum* cell, and ended up with a successfully reproducing line of *M. mycoides*; that achievement was heralded as debunking the idea that living things are "endowed with some sort of special power, force or property" (Caplan 2010).

But this claim—a welcome insight for some, perhaps a worry for others—proceeds much too quickly. As a matter of logic, the fact that a living organism has been created in a lab does not imply that it lacks whatever special power or property might otherwise have been attributed to it. If there is a God who imbues living things in swamps with that property, then He can do likewise if the thing originates in a laboratory. It helps to remember that, in a sense, scientists have been creating living organisms for a long time. Every time gametes are combined in a test tube to create embryos, life has been created. Every time animals are successfully mated, life has been created. And presumably, God (assuming He exists) has been following along all the time, doing what He likes with their efforts. What changes with synthetic biology is the method, but not the basic fact, of "creation."

A similar question would arise about human beings created through cloning, if that ever becomes feasible. Since a clone would be fully an individual being, possessing full autonomy and full moral powers and responsibilities, it is hard to see how that person would lack a soul if everyone else has one. By analogy, then, whatever special properties we find in microbial life generally could also be found in the life of a synthesized microbe. In short, the fact that human beings can create a living organism does not tell us how to understand the life that the organism possesses any more than the fact that human beings can end a life shows us how to understand it.

With this objection to synthetic biology—or with this claim about what it teaches us—it is the claim itself that makes a category mistake. Some other objections to synthetic biology accuse the field of making a category mistake. The label "playing God," for example, suggests that humans are stepping outside their proper role in the cosmos; the category mistake is a misunderstanding and violation of the category into which humans properly go. This idea is roughly the converse of the first: the idea is that because life is such a very special kind of entity, human beings dare not create it.

In a probing discussion of the ethical implications of synthetic biology, Joachim Boldt and Oliver Müller come very close to this position, if they do not actually hold it, in an argument that synthetic biology has more serious ethical implications than genetic engineering because it constitutes not merely the manipulation of nature but an act of creation. "This shift from 'manipulatio' to 'creatio ex existendo' is decisive because it involves a fundamental change in our way of approaching nature" (Boldt and Müller 2008). When we take this approach to nature, we see nature as a "blank space to be filled with whatever we wish." To see these distinctions between "fundamentally" different relationships to nature ("manipulation" versus "creatio ex existendo") as implying an ontological category mistake is perhaps to read more into Boldt and Müller's language than they intend—but neither is it obviously a misinterpretation.

The environmental philosopher Keekok Lee is more explicit. "The worrying thing about modern technology in the long run may not be that it threatens life on Earth as we know it because of its polluting effects, but that it could ultimately humanize all of nature. Nature, as 'the Other,' would be eliminated." To defend against this outcome, Lee continues, "the ontological category of the natural would have to be delineated and defended against that of the artefactual" (Lee 1999, 4).

Another way of understanding the purported category mistake—and the third way of understanding the metaphysical objection that "playing God" suggests—is to see it as a misunderstanding and violation of the category to which living things—and life in the abstract—belong. By becoming creators of life, humans are doing something inappropriate to the items in that category. In a critique concerned primarily with the social consequences of synthetic biology, the ETC Group also obliquely suggests this concern, chiefly with the use of framing language such as "Original Syn?" and "The New Biomassters" (accompanied by artwork that portrays synthetic biologists as demonic) (ETC Group 2007, 3).

There are two kinds of category mistakes, then, that might be imputed to synthetic biology, one involving an inappropriate elevation of humans and the other an inappropriate violation of life. Strictly, if there is a category gap, then it should be impossible for the science to successfully breach it. Scientists could no more trespass on God's domain than they could violate the laws of physics, and if life is ontologically distinct from matter, then science cannot alter its properties. The objection cannot be that synthetic biology achieves something humans were never meant to achieve. Humans cannot wrongly cross a metaphysical divide; they can only wrongly want to. The objection must be that by pursuing synthetic biology, we humans are representing ourselves as God.

Very likely, to understand the moral problem in either of these ways is to believe that a very serious moral problem is at hand, since it is tantamount to believing that the proper ordering of the universe, an ordering perhaps dictated by God, is under assault. Whether very many people actually have this concern about synthetic biology is not clear, however. It is not clear that people commonly think that much follows from a simple distinction between life and the rest of the cosmos. Instead, as John Evans describes (see chapter 9), they make much more of distinctions among living things. Manipulation of bacterial life, Evans speculates, lies within the range of acceptable human goals, given the category distinctions most religions and cultures recognize between humans, animals, plants, and mere objects. The Catholic Church appeared unfazed by the announcement that the genome of *M. mycoides* had been synthesized and successfully transplanted into another cell: it praised the work as "a further mark of man's great intelligence, which is God's gift enabling man to better know the created world and therefore to better order it" (BBC Monitoring Europe 2010).[1] The broad acceptance of synthetic chymosin also seems to confirm Evans's analysis. In the United States and the United Kingdom, microorganisms modified with cow genes now produce most of the chymosin used to make hard cheeses, replacing rennet, the traditional source of chymosin, which is made from the stomachs of unweaned calves. Chymosin from genetically modified microorganisms was approved by the Food and Drug Administration in 1991, with virtually no opposition.

If one nonetheless holds that the alteration of microbes reflects aspirations that run afoul of metaphysical categories, still, that position does not obviously provide solid grounds for making policy. Defending it requires appealing to and defending an account of the ontological structure of the cosmos (and showing that the ontological structure has moral implications as well). For example, one might argue that all forms of life have been divinely created, that God gave humans the ability to create life (just as he gave them the ability to destroy it), but that he proscribed the use of that ability (just as he proscribed exercising the ability to murder). Or one might argue that nature is a category of things that are and should be independent of human control (as a matter of eternal fact, not merely of human moral conventions). This is the tack taken by Lee: "the primary attribute of naturally-occurring entities is an ontological one, namely, that of independence as an ontological value" (Lee 1999, 4).

Invoking such claims as support for public policy is problematic, not only because they may not be widely shared, but also because arguably they are not good bases for public policy anyway. A full argument for this lies outside the scope of this paper, but the basic point is that, at least in the view of many political theorists, the state should adhere to some version of the liberal principle of neutrality on the kinds of views outlined above (see Jon Mandle's discussion in chapter 7 of this volume). Public policy should not endorse such views if by doing so it would exclude other views that are also compatible with the basic principles of a liberal society. Though a government could create a national park as a way of supporting citizens' beliefs about the value of nature, or support science partly in support of the intrinsic value of scientific discovery, measures that made it impossible for people who disagree with those values to carry on with their beliefs would be unacceptable. The claim that synthetic biology makes a metaphysical mistake suggests that the technology might appropriately be banned, but that would mean that the views synthetic biologists have expressed about their work would in effect be ruled impermissible. Yet those views seem to be compatible with the basic principles of a liberal society.

Moreover, any position grounded expressly on religious or metaphysical claims about the ontological structure of the cosmos may not be acceptable. A liberal government cannot build religious shrines, even if doing so does not make it impossible for some citizens to disavow that religion. It seems, then, that the metaphysical case for objecting to synthetic biology should be translated into broader, more secular moral considerations.

Moral Mistakes: The Meaning of Life

Translating the metaphysical concern into a secular moral concern can itself lead to different ways of articulating the concern. First, one could hold that synthetic biology generates a view about the human relationship to nature that conflicts with concepts that are foundational to the practice of morality itself. In this vein, Boldt and Müller argue that by presenting organisms as machinelike artifacts, synthetic biology challenges "the connection between 'life' and 'value' [and] may in the (very) long run lead to a weakening of society's respect for higher forms of life that are usually regarded as worthy of protection" (Boldt and Müller 2008, 388). Alternatively, as Boldt and Müller immediately go on to suggest, synthetic biology might change humans' conception of human agency. With the advent of synthetic biology, humans are no longer merely manipulators of nature; they can also become creators or reinventors of nature. The creation of nature might often be justifiable, they write, but also might lead to overconfidence: it "might lead to an overestimation of how well we understand nature's processes and our own needs and interests and how best to achieve them."

Concerns about denigration of life and human overconfidence are secular analogs of metaphysical concerns about violating the category of life and playing God, and some of the response to them invokes similar points. First, there is no reason to suppose that scientific investigation into how life works would force us to devalue life. The fact that an organism (be it bacterial or human) has been created in the lab does not settle what its moral value is. As Art Caplan wrote, following the announcement of a synthetic *M. mycoides*, "The value of life is not imperiled or cheapened by coming to understand how it works" (Caplan 2010).

Nor need a reductionist view of bacterial life lead to a disvaluing of higher forms of life. One study of public attitudes toward synthetic biology suggests that people are not nearly as bothered about creating and modifying single-celled organisms as they would be if synthetic biology took on the creation and modification of higher-level organisms (Royal Academy of Engineering 2009). If we are not guided by a single large distinction between life and matter, but by a series of distinctions among living and inanimate things, then attitudes toward bacterial life may be connected only remotely to attitudes toward human beings. Certainly, people have sometimes viewed animals as machinelike and morally insignificant without that view having any diminishing effect on views about human life. What has been held special about human life is often not the merely biological fact of being alive, which we share with other forms of life, but an assortment of cognitive and emotional capabilities that seem to distinguish humans from animals and other living things—prominently, the capacity to have moral values and the capacity to reason. And since the capacity to reason gives us capacities to understand nature, to master it, and to rework it, synthetic biology might even be seen as reminding us of the value of human life. In effect, this is the other side of the claim that humans might be playing God; maybe we will see ourselves as gods. (And in keeping with the epigram that you are what you pretend to be, maybe we will act more like gods. Or not!)

That we might see ourselves as gods is tantamount to Boldt and Müller's second concern, about human overconfidence. If the first point is that synthetic biology might lead us to underestimate life, the second is that it might lead us to overestimate ourselves. In particular, we might overestimate our capacity to understand and helpfully change the world. Strictly, this prospect does not seem to be an ethical problem in its own right; it does not pose the possibility of undermining the practice of morality in the same way that altering the concept of life purportedly might. The point is rather a more straightforward consequentialist one. The point seems to be that, if we get really good at this stuff, we might lose our moral moorings and use our abilities recklessly—and end up doing work that has unintended bad effects. Of course, this possibility is always there; we do not need to excel at synthetic biology in order to do things that have unintended bad effects. Our very first project might have unintended bad effects. In any event, once we see that the concern is (merely) about consequences, it is detached from any special arguments about how synthetic biology might change our systems of meaning.

How synthetic biology might change our moral concepts might turn out to depend partly on where synthetic biology happens. If the BioBricks vision really pans out and high-school students working at kitchen tables are someday creating new organisms, then synthetic biology might well seem to impart a widespread human capability, and perhaps its effect on our view of life and humanity would likelier be significant. But if it were restricted to the laboratory and the factory, its use monitored by regulatory authorities in some fashion, and the release of organisms into the wild forbidden or restricted, then it might be seen as a rather exceptional power—something that must be guarded and restricted and that does not broadly change the human relationship with life.

Finally, one might wonder why synthetic biology's implications for concepts constitutes an objection. It would not be the first time that science has challenged our views about life and our role in the cosmos. Whenever science challenges the factual claims of a religious text from which people also derive moral guidance, it has implications for moral concepts. The Copernican revolution removed humans from the center of the cosmos. The Darwinian revolution rubbed out a clear distinction between humans and other animals. In the nineteenth century, when the German chemist Friedrich Wöhler invented synthetic urea, he established that whatever special property is possessed by living things (if there is one) is not causally necessary for the production of organic chemicals. It could be only a metaphysical property, not also a physical property—a kind of "vital force"—that could be isolated in the lab. But were any of these scientific developments therefore wrong? From a modern perspective, they look more like moral progress than moral decline.

Preservation of Nature

Another way of arguing that synthetic biology raises a moral concern about the human relationship to nature is to argue that it is in some way damaging to nature. The touchstone for this sort of concern is the common environmentalist belief that the environment should be protected not just to make sure that the environment is beneficial for humans, but also to protect the environment from humans. This kind of concern encompasses endangered species, untrammeled wildernesses, "untamed rivers," old-growth forests, and mountains, canyons, and caves. We should approach these things, many feel, with a kind of reverence or gratitude.

The environmental philosopher Christopher Preston has objected to synthetic biology along these lines, by arguing that synthetic biology violates the distinction between natural and artifactual in a way that "traditional" molecular biotechnology does not. "The relevant difference is that traditional biotechnology has always started with the genome of an existing organism and modified it by deleting or adding genes" (Preston 2008). By contrast, synthetic biology aims "to create an entirely new organism," and it thereby crosses a line that, to environmentalists, is basic and cherished: it "departs from the fundamental principle of Darwinian evolution, namely, descent through modification."

Preston's objection to synthetic biology echoes the science writer Michael Pollan's explanation for why he found genetically modified potatoes troubling. When new varieties are created through conventional breeding, he argued, they can be seen both as products of human creativity and as an evolutionary adaptation of the plant's; there's a "plant's eye view" of the emergence of the new variety, as Pollan puts it. Neither account of their emergence is complete. To fully explain a new variety, one must talk both about the genetic lottery and about human selection. "The plant in its wildness proposes new qualities, and then man... selects which of those qualities will survive and prosper" (Pollan 2001, 196). With genetically modified foods, however, the story of human creativity is a full account, and the evolutionary story—the plant's perspective—drops out.

But what to make of this? Pollan's account of genetically modified foods helps illuminate the limits of Preston's evolutionary objection to synthetic biology. Synthetic organisms, like genetically modified potatoes, lack an evolutionary story, and therefore lack whatever value is found in things to which that story belongs, but they do not constitute a threat to that story—they do not deprive us of the value attached to that story—unless they get out of hand and become an environmental hazard. (Christopher Preston, in chapter 6 of this volume, and Mark Bedau and Ben Larson in chapter 4, expand on objections to Preston's objection.) Pollan did not draw the conclusion that it was wrong to create genetically modified potatoes. He just didn't eat them.

The human-nature issues that have most alarmed the public, and that have led to public policy, have concerned quantifiable damage to the natural world, and especially permanent damage: when the passenger pigeon was killed off, for example, it was gone forever. When chestnut blight was introduced to North America, the American chestnut lineage collapsed, substantially changing the forests of eastern North America in the process. But the creation of an organism—the aim of synthetic biology—is not necessarily damaging. At least if the organism remained confined to the laboratory or processing plant, the natural world around us would remain unperturbed. Possibly, if the enthusiasts' hopes are well founded, the creation of synthetic organisms will even turn out to be environmentally beneficial. Possibly, synthetic biology offers one route to a kind of environmental restoration: The only possible way of bringing back the passenger pigeon would be to engage in an avian version of the work with *M. mycoides*: to synthesize its genome in a suitable egg. And since a few American chestnuts survive, the Asian chestnut genes that confer resistance to chestnut blight could be introduced to American chestnuts via genetic engineering (as well as through crossing and back-crossing).

One rejoinder to this line of thought might be that, though synthetic biology is not itself necessarily damaging to nature, it evinces and encourages a general view of the relationship of humans to nature that is damaging. It fosters a view that nature can be adjusted to human wants and needs, rather than that humans must accommodate nature.[2] It is not necessary to reduce our fuel consumption (one might be led to think); we can just twiddle with these organisms to devise a brilliant new way of simultaneously producing fuel and absorbing the carbon dioxide added to the atmosphere when the fuel is burned.

If synthetic biology were as potent as that—if it could single-handedly solve our environmental problems, this would be a very strange objection. It would have us, out of love of nature, disavowing a tool that could protect nature. More likely it offers at best a partial solution to those problems. Even so, there is an inherent tension within the position. To generate an argument against conducting synthetic biology, the point must be that it ultimately only makes the problem worse—somewhat in the way that adding a lane to a highway to reduce traffic congestion often ultimately encourages still heavier traffic and produces even more congestion. This may be right, but the reasoning needs to be spelled out and defended a little more.

In fact, the impetus to protect the environment seems unchanged by synthetic biology. There are many environmental problems that synthetic biology does not appear to touch—the threat to sea turtles from shrimp fishing, the draining of wetlands to build malls, the effect of light pollution on migrating birds—and those it does touch are too great to be solved by any one measure. The problems posed by the production and consumption of oil-based fuels are so vast and so serious that, even if we find new ways of producing fuels, we still need to reduce fuel consumption. In short, given the facts of the environmental crisis, even if we pursue synthetic biology, we also still need to alter human behavior to accommodate natural realities, and we should not suppose that the technology lets us think otherwise. There is no general choice to make between adjusting nature to human behavior and adjusting human behavior to nature.

Moreover, if synthetic biology is understood as a phenomenon of the laboratory and the factory—as another way of producing things that otherwise come out of oil refineries—then it is hard to see that deciding to go forward with synthetic biology would always reflect a preference for altering nature to suit human demands. Most applications so far envisioned for synthetic biology are in the context of human practices that are not themselves very expressive of human reverence for nature. Synthetic biology is different in this regard from the genetic modification of crops and livestock. Agriculture and cuisine are human practices that many people associate with special ideals about the human relationship to nature. But the most promising applications for synthetic biology are so far in drug development, fuel production, and other industrial applications. One may want to drive a car that does comparatively little environmental damage, but car-driving is not itself an activity that reflects a connection to the earth or a conversational relationship with life. With agriculture and cuisine, engaging in the activity could be expressive of a certain view of the human relationship to nature; one might go into the garden—to plant or to harvest—precisely to honor one's sense of a connection to the earth. With car-driving, it is restraint that expresses a sense of connection to nature.

Synthetic Biology and Nature

Synthetic biology seems, in some ways, to be an archetypal case of human alteration of nature: what could be more significant than the synthesis of living organisms to serve human needs? At the same time, in its current form, synthetic biology does not really seem to change the terms of the game. It touches a nerve—it raises a general concern about the human-nature relationship that is important and deep—but when we investigate more closely and try to articulate just how synthetic biology might change that relationship, the conversation seems to spin out into inconclusive eddies. Microbes engineered for industrial use (and rendered safe for humans and the environment)? Why not?

Some concerns about synthetic biology are simply not well founded: it does not show that life is nothing but a chemical process, for example, nor is it likely to lead people to undervalue life. Other concerns seem better founded, but still do not make a large moral or policy difference. Concerns grounded on metaphysical positions do not translate into policy positions very easily, for example. Concerns about its consequences for nature are mirrored by concerns about the consequences for human health and well-being, which typically are less problematic bases for public policy making. And finally, as a way of changing nature, it is not obviously worse than many other things humans already do, and that environmentalists regard as intrinsically acceptable. Synthetic biology in its current form is not necessarily damaging to nature, and it does not even necessarily alter the terms of the human relationship to nature. It does not necessarily change either the world around us, or human practices themselves, in ways that deprive us of something intrinsically valuable.

To be sure, the technology could take some other form, in some new context, and end up depriving us of something we value. One way this might happen is if synthetic biology were to become possible for more complex organisms, or even for human beings. Also, the technology might end up depriving us of something we value if it were to be used in ways that significantly altered the environment. The threats of bioterrorism and "bioerrorism" are worth taking seriously for the same reasons we take seriously the environmental threats posed by other kinds of industry or agriculture: not only do they pose a risk to human health and well-being, they also may threaten naturally occurring species and ecosystems. The environmental risk raises both consequentialist concerns and concerns related to the intrinsic value widely attributed to nature.

A straightforward concern that synthetic organisms might damage the natural world appears to be as plausible a candidate for grounding public policy as objections about the environmental damage caused by other human activities. It requires no special defense beyond that already offered for policies to protect rare species and undeveloped lands. For most uses involving contained microorganisms, these concerns seem to run in tandem with consequentialist concerns, and since the latter are already quite serious, the former probably do not generate any policy considerations not already generated by the latter. But this could easily change if synthesized microbes were developed expressly for environmental uses—for cleaning up oil spills or rebuilding ocean food chains, for example. Those applications might benefit humans while still posing risks to the environment—to endangered species, for example—and then the intrinsic value given to nature would be highly relevant. And perhaps there would simultaneously be environmental benefits as well. A balance would have to be struck, just as with other kinds of environmental concerns: it is surely sometimes appropriate, all things considered, to cut down trees, mine for ore, drill for oil, hunt and fish, and so on.

For anyone who cares morally about the human relationship to nature, the prospect of synthesizing living organisms has to be eyebrow-raising. But one must still try to assess that reaction—articulate it, explain it, and defend it. Of course, we should not expect to be able to defend our moral reactions all the way down. At the same time, as I have also argued, we should expect that at the end of the day, after we have evaluated our initial reactions, not all of them will be left standing. We will have changed our minds about some. It need not be surprising or troubling, then, to find that in defending moral concerns about nature we end up shrugging off concerns about synthetic biology.

Notes

1. Quoted in Presidential Commission for the Study of Bioethical Issues, New Directions: The Ethics of Synthetic Biology and Emerging Technologies (Washington, DC: Presidential Commission for the Study of Bioethical Issues, 2010), 138.
2. I owe this thought to Bruce Jennings.

References

BBC Monitoring Europe. 2010. Vatican dismisses synthetic cell's life-giving dimensions, lauds science research. May 25.

Boldt, Joachim, and Oliver Müller. 2008. Newtons of the leaves of grass. *Nature Biotechnology* 26:387–389.

Caplan, Art. 2010. Now ain't that special? The implications of creating the first synthetic bacteria. Scientific American online Guest Blog, May 20. http://blogs.scientificamerican.com/guest-blog/2010/05/20/now-aint-that-special-the-implications-of-creating-the-first-synthetic-bacteria/.

Cho, Mildred K., David Magnus, Arthur L. Caplan, and Daniel McGee, and the Ethics of Genomics Group. 1999. Ethical considerations in synthesizing a minimal genome. *Science* 286 (5447):2087–2090.

Edge: The Third Culture. 2008. Engineering biology: A talk with Drew Endy. Edge 237. http://www.edge.org/documents/archive/edge237.html.

Group, E. T. C. 2007. *Extreme Genetic Engineering: An Introduction to Synthetic Biology*. Ottawa: ETC Group.

Lee, Keekok. 1999. *The Natural and the Artefactual: The Implications of Deep Science and Deep Technology for Environmental Philosophy*. Oxford, UK: Lexington Books.

Pollan, Michael. 2001. *The Botany of Desire: A Plant's Eye View of the World*. New York: Random House.

Presidential Commission for the Study of Bioethical Issues. 2010. *New Directions: The Ethics of Synthetic Biology and Emerging Technologies*. Washington, DC: Presidential Commission for the Study of Bioethical Issues.

Preston, Christopher J. 2008. Synthetic biology: Drawing a line in Darwin's sand. *Environmental Values* 17:23–39.

Regis, Ed. 2008. *What Is Life?* New York: Farrar, Straus, and Giroux.

Royal Academy of Engineering. 2009. *Synthetic Biology: Public Dialogue on Synthetic Biology*. London: Royal Academy of Engineering.

Wade, Nicholas. 2010. A decade later, genetic map yields few new cures. New York Times, June 12.

II

The Value of Synthetic Organisms

4

Lessons from Environmental Ethics about the Intrinsic Value of Synthetic Life

Mark A. Bedau and Ben T. Larson

Synthetic biology is the attempt to "engineer complex artificial biological systems to investigate natural biological phenomena and for a variety of applications" (Andrianantoandro et al. 2006; see also Endy 2005). We will use the expression "synthetic life-forms" to refer to the different kinds of synthetic organisms produced in synthetic biology laboratories. These organisms today are typically various kinds of genetically modified bacteria. Even if most (or even all) of the material in a synthetic life-form comes from natural forms of life, we still consider it to be "synthetic" because it is produced through the intentional activity of laboratory scientists. Synthetic life-forms are synthetic to various degrees. We are concerned here only with those synthetic life-forms that have undergone massive human manipulation.

Much of the best-known work in synthetic biology is "top down" in the sense that it starts with some pre-existing natural living system and then reengineers it for some desired purpose (e.g., Ro et al. 2006), perhaps by synthesizing (Gibson et al. 2008) or transplanting (Lartigue et al. 2007) entire genomes. Another approach to engineering novel biological systems works strictly from the "bottom up" in the sense that it attempts to make new simple kinds of minimal chemical cellular life, using as raw ingredients only materials that were never alive (Szostak et al. 2001, Rasmussen et al. 2004); these bottom-up creations are sometimes called "protocells" (Rasmussen, Bedau, Chen, et al. 2009).

The most notable recent achievement in top-down synthetic biology is the so-called synthetic cell created by the JCVI team (Gibson et al. 2010). About 1 percent of the dry weight of the synthetic cell is synthetic: the artificially synthesized copy of the natural genome of *Mycoplasma mycoides*. The artificial *M. mycoides* genome was transplanted into a natural *Mycoplasma capricolum* with a deactivated natural genome. The 1 percent artificial and 99 percent natural hybrid system expressed its new *M. mycoides* genome and became an *M. mycoides* cell.

Since synthetic biology promises to transform biotechnology (Endy 2005, Carlson 2010), the topic is increasingly visible in the popular press (e.g., Specter 2009). Work on the social and ethical implications of synthetic biology has exploded and is producing a steady stream of papers (e.g., Cho et al. 1999; Boldt and Müller 2008; Bedau et al. 2009; Kaebnick 2009), books (e.g., Schmidt et al. 2010; Bedau and Parke 2009), journal issues (e.g., Schmidt 2009), commentary (e.g., Bedau et al. 2010), and institutional reports (e.g., Garfinkel et al. 2007; van Est et al. 2007). This sustained ethical reflection is having an increasingly direct effect on discussion of social policy (e.g., Kaebnick 2009; Presidential Commission for the Study of Bioethical Issues 2010). Most of the broader discussions of synthetic biology address its aggregate social utility, including such considerations as economic benefits and threats, and safety and security concerns (e.g., Garfinkel et al. 2007). In contrast, we focus on whether the synthetic life-forms created in synthetic biology have any intrinsic value that deserves moral consideration.

Parallel questions about the intrinsic value of animals and the environment have been addressed for many years in environmental ethics (e.g., Callicott 1989; Katz 1992; O'Neill 1993; Sandler 2007; Nolt 2009). Like synthetic biology, environmental ethics discusses the intrinsic value of nonhuman forms of life (Rolston 1982; Taylor 1986; Elliot 1992) and the normative significance of the natural as opposed to the artificial (Leopold 1949; Hettinger and Throop 1999). In fact, environmental ethics contains the most extensive discussion of intrinsic values of nonhuman life-forms.

One question, perhaps the threshold question, raised by synthetic biology is whether the synthesis of organisms runs afoul of the intrinsic value of nature. The earlier chapters in this volume give some attention to this question. In this chapter, we use the discussion of intrinsic value in environmental ethics as an opportunity both to reconsider that question and to pivot to a second question, namely, whether synthetic life-forms might themselves turn out to have intrinsic value—a question that the following two chapters will take up as well. We follow Sandler (2010) and distinguish three types of intrinsic values—intrinsic subjective value, intrinsic objective value, and inherent worth. It turns out that the environmental ethics literature attributes all three kinds of intrinsic values to types of nonhuman life-forms. We conclude that parallel thinking should lead us to attribute analogous intrinsic values to synthetic life-forms.[1]

Escaping the Conditional Fallacy

We start with Preston's pioneering application of Leopoldian environmental ethics to synthetic biology. Preston asked how considerations from environmental ethics might be used to criticize synthetic biology. Preston's argument focuses on the contrast between the natural evolutionary history of natural forms of life, and the unnatural, artificial, human-engineered process of synthesis in the laboratory by which life-forms are created by synthetic biology. The heart of the argument is captured in the following passage:

> If, like Leopold, you are an environmentalist who puts normative stock in the idea of the historical evolutionary process then synthetic biology should be opposed on deontological grounds due to the way it disconnects the biological artefact from this evolutionary history. This departs from the natural evolutionary process in the way that is different from any previous supercession in biotechnology. If that natural evolutionary process has substantial normative significance then these biotic artefacts are morally different from all previous ones. The biotic artefacts produced by synthetic biology depart from nature in a more radical way than anything that has come before. (Preston 2008, 36)

We generally agree with Preston's premise about the novelty of synthetic biology. Contemporary synthetic biology, especially bottom-up synthetic biology or protocells, enables human engineers to design and construct synthetic life-forms intentionally, consciously, and rationally, with a flexibility and precision of control that is without previous comparison. So the construction process by which synthetic life is created can differ radically from the natural evolutionary process that produced all natural forms of life. And we agree with Preston that the creations of synthetic biology "depart from nature in a more radical way than anything that has come before." At the very least, this is reason enough to give the creation of synthetic life independent evaluation.

In addition to the novelty premise, Preston's argument also invokes a Leopoldian premise about the normative significance of a natural evolutionary history. Putting the two premises together yields the following simple overall argument:

Premise 1 If a life-form has a natural evolutionary history, then it has intrinsic value.

Premise 2 Synthetic life-forms have no natural evolutionary history.

Conclusion Synthetic life-forms have no intrinsic value.

When Preston talks about evolutionary histories having normative significance, we assume this means that creatures derive normative significance from their natural evolutionary histories; more specifically, their evolutionary histories give them some kind of genuine intrinsic value, and so they deserve our moral consideration and respect. So, a larger issue raised by Preston's argument is the purported intrinsic value of natural life-forms.[2]

We do not question the premises in Preston's argument. We view Premise 2 as a simple matter of empirical fact.[3] Premise 1 is different; it expresses the Leopoldian value of having a natural evolutionary history, a value widely shared in environmental ethics (e.g., Rolston 1982; Katz 1992; Sober 1986; Light 2000).[4] Although someone could question why having an evolutionary history would make something intrinsically valuable, here we will simply assume Premise 1 for the sake of argument and ask what would follow.

Premise 1 describes one particular kind of connection between an organism's intrinsic value and its evolutionary history. Some other kinds of connections are also plausible. We will explore them in a moment.

We also do not question the ultimate significance of the argument's conclusion. If synthetic life-forms lack intrinsic value, what follows about what anyone should do? Should regulations and policies affecting synthetic biology be created or modified? Should we change the way research and development in synthetic biology is funded or regulated? Should some research in synthetic biology be banned? Who should have legal oversight of synthetic life-forms, and should oversight be voluntary or compulsory? We mostly set aside these questions for another day. Our focus at the moment is the logic that connects environmental premises with conclusions about the intrinsic value of synthetic life-forms.

Once we turn to the logic of the argument, we see that the argument is an instance of the conditional fallacy, and so is invalid.[5] The conditional fallacy arises in this argument because something can have intrinsic value even if it lacks the intrinsic value produced by a natural evolutionary history. It might have some *other* kind of intrinsic value. Thus, the argument is implausible unless it is reformulated to avoid the conditional fallacy.

One trivial way to avoid the fallacy is to replace Premise 1 with its contrapositive: If an organism has intrinsic value, then it has a natural evolutionary history. The contrapositive makes the argument valid; if having a natural evolutionary history is a necessary condition for having intrinsic value, then of course synthetic life-forms lack value since they lack a natural evolutionary history. But the contrapositive is implausible.

Even if we grant that having an evolutionary history is *one* way to have intrinsic value, there is no reason to conclude it is the *only* way.

A more plausible but more complicated way to avoid the conditional fallacy is to replace Premise 1 with an exhaustive list of all of the possible kinds of intrinsic value, and modify Premise 2 to claim that the biotic artifacts produced by synthetic biology have none of the properties on the list. These two premises entail that no synthetic life-form has any kind of intrinsic value. If synthetic life-forms lack all of the possible kinds of intrinsic value, then the original conclusion stands; synthetic life-forms have no intrinsic value.

Constructing an exhaustive list of all of the possible kinds of intrinsic value involves surveying all plausible candidates. This is where the environmental ethics literature is helpful. For nothing has expanded our appreciation for the intrinsic value of all nonhuman forms of life and of the natural world more than environmental ethics.

Intrinsic Values in Environmental Ethics

The environmental ethics literature contains appeals to many kinds of intrinsic value. Table 4.1 lists the kinds of intrinsic value referenced in some representative discussions of environmental ethics: Rolston's classic, comprehensive discussion (Rolston 1988) and one of the many excellent recent anthologies (Schmidtz and Willott 2002). We treat table 4.1 as an approximation of an exhaustive list of all possible kinds of intrinsic value, and ask whether these values are missing in synthetic life-forms. Our framework for categorizing intrinsic values comes from Sandler (2010), who distinguishes three kinds of intrinsic value.[6] *Intrinsic subjective value* is the value that something possesses in virtue of someone's valuing it for what it is, rather than for its usefulness as a means to an end. *Intrinsic objective value* is the value that something possesses in and of itself, independent of whether anyone actually values it. *Inherent worth* is the value that something possesses in virtue of having a good of its own (or interests) that valuers (or moral agents) ought to care about.

Table 4.1 groups different kinds of intrinsic value by row. Each row represents a certain kind of value, which might be describable with multiple words or phrases. The three sections of the table correspond to Sandler's three kinds of intrinsic value. Intrinsic subjective values are relational values that depend on someone's subjective opinion.[7] The defining feature of this group is that the intrinsic value exists only because of the existence of the right relationship with someone's subjective acts of attributing intrinsic value. Examples include someone valuing a painting because of its aesthetic beauty, or valuing a natural landscape because it inspires awe. It also includes someone negatively valuing or *dis*valuing something that threatens to undermine a central presupposition of his or her culture (e.g., the existence of a god or the mechanistic chemical quality of human life).

Intrinsic objective values, the second category in table 4.1, do not depend on anyone's subjective opinions. They are systemic or structural properties that characterize certain kinds of complex systems that have many interacting parts; the systemic properties concern the organization and operation of the whole system and all its parts. Examples of systemic properties that seem ripe for this kind of system-level analysis include integrity and stability, complexity, diversity, self-regulation; resilience and self-repair; autonomy and creativity; even negative entropy. This group also includes the intrinsic objective value attributed to natural and wild landscapes and communities and their contingent natural evolutionary histories that Preston's argument invokes.

Inherent worth is the third group of intrinsic values in table 4.1. Inherent worth is possessed by things (typically, certain forms of life) that have their own interests, purposes, cares, or biological needs, and that strive to remain healthy or avoid injury. The environmental ethics literature is famous for expanding the scope of the interests of nonhuman life-forms that deserve moral consideration. In addition, the animal liberation literature (Singer 1975) has raised awareness of the interests of all sentient creatures. A number of environmentalists think we should extend moral consideration to all forms of life, even invertebrates and plants, because all forms of life have interests (this idea is discussed in the next section). One could even attribute interests to whole biotic communities or landscapes. All of these kinds of interests ground the selected examples of inherent worth.

Intrinsic subjective values play an important role in society at large. Society supports the arts and music; it preserves wilderness areas and historical sites, and it protects endangered and indigenous species. Sometimes intrinsic subjective values are thought to complement intrinsic objective value and inherent worth, as when Leopold (1949) and Rolston (1988) admire the beauty and sublimity of nature and life. Elsewhere intrinsic subjective values appear as alternatives to the other types of intrinsic values. Social or cultural values are emphasized by Sagoff (1997) and Bookchin (1988). Russow (1981) and Sober (1986) champion the importance of aesthetic considerations in environmental ethics. Sober

Table 4.1
Intrinsic values in environmental ethics

Intrinsic subjective value	
Beauty, harmony, sublimity, complexity, rareness, profundity, authenticity	Russow (1981); Rolston (1988); Leopold (1949); Sober (1986); Krieger (1973); Callicott (1989)
Cultural importance, social importance	Bookchin (1988)
Intrinsic objective value	
Autonomy, self-regulation, spontaneity; spontaneously evaluative, emergent	Rolston (1988); Taylor (1981); Katz (1992); Norton (1991)
Complexity, diversity	Devall and Sessions (1985)
Constructivity, creativity, negentropy	Rolston (1988); Norton (1991)
Naturalness (nonartificiality), wilderness	Rolston (1988); Katz (1992); Leopold (1949); Krieger (1973); Light (2000)
Self-defense, self-regulation, autonomy	Rolston (1988); Taylor (1981); Katz (1992)
Resiliency	Rolston (1988); Hargrove (1989)
History: long, contingent, natural, evolving	Sagoff (1997); Rolston (1988)
Inherent worth	
Personal interests, purposeful behaviors, strivings, cares; self-defense	Rolston (1988); Attfield (1981); Varner (2002); Feinberg (1974); Taylor (1981); Katz (1992)
Biological needs	Varner (2002)
Health, the capacity to be harmed, to have things be good and bad, to be injured and to receive benefits, to suffer, to care, to have personal interests	Rolston (1988); Midgley (1983); Stone (1972)
Membership in a (biotic) community that has interests, and that has parts that are interdependent, such that the health of the parts requires the health of the whole	Taylor (1981); Midgley (1983); Light (2000); Katz (1992)

Note: Entries as reflected in a classic survey (Rolston 1988) and a recent anthology (Schmidtz and Willott 2002). Each row represents a relatively unified category of intrinsic value. Following Sandler (2010) we group each case under intrinsic subjective value, intrinsic objective value, and inherent worth.

notes that we value authenticity and rarity in natural systems just as we value such qualities in art.

One example of an appeal to intrinsic objective value is Rolston's argument that the network of interactions by which living systems organize and thereby resist entropy is a source of "natural value" (1988, 19). Ecosystems are organized, generative, and creative. Selective pressures in the ecosystem generate and sustain biodiversity, and create ties within the biotic community that contribute to the dynamic stability of the whole ecosystem and all the life-forms in it. Norton also advocates creativity in nature as a source of value (Norton 1991). For Rolston, Norton, Taylor, and Katz, the dynamic stability and autonomy of both organisms and ecosystems are valuable and ought to be respected. Ecosystems not only generate diversity; they are sustained by it. So the spontaneous and autonomous functioning of ecosystems serves the interest of the ecosystem, over and above the interests of the individual organisms in the ecosystem. Another property thought to have intrinsic objective value is the property of being wild or naturally evolved. Rolston, Katz, Light, Leopold, and Krieger (1973) all agree that wildness is valuable. Krieger, Leopold, and Rolston hold that wilderness is valuable to experience, and Katz and Light hold that the wild has a "moral claim" on humanity (Katz 1992, 176). Krieger has it that wilderness can be thought of as a "state of mind" (1973, 162) and a source of revitalization as well as a biological system organized in a particular way. For Katz, the moral standing of wildlife is the result of our being part of a moral community with other living things in the natural world. Another property linked to wildness is evolutionary history.

Finally comes inherent worth. Many have identified a kind of organismal value possessed by any living thing (Rolston 1988; Attfield 1981; Taylor 1981; Katz 1992; Varner 2002; Midgley 1983; Stone 1972). These authors agree that living things act purposefully and have self-interest. Rolston argues that this kind of self-interest is value-indicating in that an organism protects the "good-of-its-kind" as expressed in its genetic code (1988, 109). Many agree with Rolston, including Attfield, Varner, and Taylor. Varner and Attfield in particular argue that biologically based needs count as interests. Taylor and Rolston hold that interests are entailed by the purposeful behavior of an organism, as it pursues a good of its own. Another indication of having interests or rights, and hence intrinsic value, is being able to be harmed or benefited, a perspective advocated by Rolston, Regan, Midgley, and Stone. Feinberg (1974) and Singer (1975) disagree with this permissive interpretation of interests; they argue that sentient desires are necessary for interests, and so interests must be psychological. Foreseeable synthetic life-forms would have inherent worth only if Feinberg and Singer are wrong and interests need not be psychological.

Intrinsic Values of Synthetic Life-forms

We can now answer whether the products of synthetic biology have intrinsic subjective value, intrinsic objective value, or inherent worth, and if so, with what consequences. In the process, we will assume that in the foreseeable future synthetic life-forms will be very roughly equivalent in size and complexity to the simplest natural forms of life, such as bacteria.

Intrinsic subjective values are relational; they depend on the subjective and contingent psychological attitudes of people who take things to be beautiful, culturally important, or a source of wonder worthy of scientific investigation. Many intrinsic subjective values are widely given great normative significance. For example, contemporary society generally cherishes and celebrates important works of art, and devotes substantial resources to preserving and displaying them in public museums and performances. Many other intrinsic subjective interests are highly idiosyncratic and personal, and not widely shared. By encouraging everyone to appreciate the intrinsic subjective value in other forms of life and of the environment, environmental ethics promotes a pluralism of intrinsic subjective values.

Because these values are subjective, almost anything can be valued for what it is by someone or other. Since people can subjectively value anything, we should expect synthetic life-forms to be subjectively valued and disvalued by various people in various ways. Thus, there is no doubt that synthetic life-forms can have intrinsic subjective value. Someone will find them beautiful, culturally or historically important, or the like. But this subjective kind of intrinsic value may have no more moral or political importance than any other subjective value. The subjectivity of intrinsic subjective values caps their ethical significance.

A special intrinsic subjective value of synthetic biology comes from its promise to illuminate many of the remaining molecular mysteries of minimal chemical life. This is not an intrinsic value of synthetic life-forms themselves, but an intrinsic value of our *scientific knowledge about* synthetic life-forms. The knowledge created by synthetic biology is valued, of course, by scientists working in the field. Much of this knowledge has pragmatic commercial value, but that does not prevent many scientists from valuing the new information for its own sake. The open-ended combinatorial reprogramming of life provided by synthetic biology opens the door to an epistemological revolution. It gives us unprecedented power to investigate the empirical possibilities of minimal chemical life. There is great intrinsic epistemic value in the knowledge that can be produced by this new epistemic tool.

For a certain segment of society, the knowledge produced by synthetic biology will have a great intrinsic subjective *disvalue*. This disvalue comes from the culture shock created when synthetic biology disrupts our picture of our place in the cosmos. Specifically, the advancement of synthetic biology emphasizes that simple life-forms are nothing more than complicated biochemical mechanisms. And if that is true for simple life-forms, it is presumably true for all the other life-forms that evolved from them, including humans. To many people, the mechanistic implications of synthetic biology deflate human freedom and morality (Cho et al. 1999). In our view, the ultimate consequences of the mechanistic view of all life remain an open question. It might be a relatively direct logical consequence of synthetic biology that all forms of life are just extremely complicated and massively interconnected molecular mechanisms; this perspective was already widespread among scientists before synthetic biology, but synthetic biology gives it new and vivid emphasis. However, any conclusions about whether this mechanistic perspective necessarily robs humans of freedom or morality are much more tenuous. The larger implications are that, as synthetic biology unfolds, we should be sensitive to the culture shock it might cause, and we should keep an open mind about its larger implications.

In addition to intrinsic subjective values, the environmental ethics literature also appeals to a number of objective systemic properties as candidates for intrinsic objective values. We noted above that these properties include autonomy, resilience, self-regulation, stability, self-organization, diversity, and complexity. Candidates for intrinsic objective values also include being wild and having a natural evolutionary history. The whole class of intrinsic objective values is rather controversial, and we will leave their defense to the environmental ethics literature. Instead, we simply ask what would follow from those values.

Preston's argument can be construed as claiming that synthetic life-forms lack a significant kind of intrinsic objective value because they lack a normal evolutionary history. A related complaint devalues synthetic life-forms because they are artificial and so lack the intrinsic objective value of the wild. On the other hand, it is striking that synthetic life-forms actually possess many of the objective systemic properties listed in table 4.1.

For example, laboratory populations of protocells already exhibit various forms of autonomy, self-regulation, self-defense, spontaneity, and creativity (Rasmussen, Bedau, Chen, et al. 2009). The exemplification of these properties in synthetic life-forms will only become stronger and more robust as synthetic biology progresses and synthetic life-forms become increasingly interwoven in a complex and diverse ecology. Therefore, synthetic life possesses (or soon will possess) many of the objective systemic properties that environmental ethics associates with intrinsic objective value.[8] So, many of the intrinsic objective values from environmental ethics actually *support* the normative significance of synthetic life, because they apply to *all* forms of life.

The property of having a natural evolutionary history deserves further discussion. Here is one reason why having a natural evolutionary history could imbue something with intrinsic objective value. Natural evolutionary histories are produced by a special kind of process, one that tends to produce things that reflect a certain "wisdom of nature"; that is, natural evolutionary processes tend to produce well-adapted and well-regulated organisms and biotic communities. This natural wisdom of well-designed systems is an intrinsic objective value. Of course, the wisdom of nature is limited and works only in certain contexts. Natural forms of life contain many frozen accidents and suboptimal designs, such as the panda's thumb (Gould 1992).

We implied earlier that the value of a natural evolutionary history is debarred from synthetic life-forms, for they have no natural evolutionary history. However, there is more to this story. One complication is that some synthetic life-forms are produced using processes such as directed evolution (Yokobayashi et al. 2002) or evolutionary design of experiments (Caschera et al. 2010) that embody and exploit a certain kind of "wisdom of nature." Directed evolution and evolutionary design of experiments both mimic the power of evolutionary search and design processes, which occurs most notably in adaptive Darwinian evolution driven by natural selection. Darwinian evolution by natural selection is a source of the wisdom of nature, and both directed evolution and evolutionary design of experiments tap into that source. This suggests that synthetic life-forms could have the intrinsic objective value (if any) that comes from the wisdom of nature. Being created in the laboratories of teams of human scientists makes synthetic life-forms unlike any natural life-form. But they can still have many of the valuable features of natural life-forms, especially if their production is carefully designed and directed so as to create an autonomous, adaptive, evolutionary process. Human scientists create the process, but they then let that process run autonomously. The evolutionary histories that result are stochastic adaptive responses in populations, just like any other evolutionary history. So, the intrinsic objective value of a natural evolutionary history might apply to certain synthetic life-forms. It depends on exactly which kind of evolutionary history is needed and why.

Furthermore, if synthetic life-forms were released into the environment and left to fend for themselves, they could adapt, mutate, and evolve in ways that no one anticipates today. This kind of evolutionary process happens with existing microbial life, and there is every reason to think it would also happen with synthetic life-forms. As this process continues over time, the synthetic life-forms could build up and come to possess an evolutionary history. Descendants could accumulate a more and more detailed contingent evolutionary history and start to grow "wild." Accordingly, descendants of synthetic life-forms would come to possess the sort of intrinsic values possessed by wild and naturally evolved organisms. In time the offspring of synthetic life-forms could become integral parts of landscapes every bit as wild as the Wisconsin sand counties that Leopold loved. In this way, the descendants of synthetic life-forms would come to possess any intrinsic objective value bestowed by a natural evolutionary history.

We noted above that environmental ethics champions the idea that all organisms have an inherent worth that deserves our respect and that creates certain obligations for us. The reason for attributing inherent worth to all forms of life is that all forms of life have biological interests, needs, purposes, drives, and the like. It is easy to understand the appeal of this idea. All organisms have their own interests and strive to achieve their own goals and purposes. These interests derive from the organisms' biological needs.[9] It follows that synthetic life-forms also have inherent worth, because they are just as alive as any other form of life and have their own biological interests, needs, and the like. So, environmental ethics provides a plausible case for attributing inherent worth to synthetic life-forms. This implication of environmental ethics might be easy to miss, for many authors (e.g., Rolston 1982; Katz 1992) claim that technology is unnatural and that artifacts have no interests of their own. Synthetic forms of life directly challenge this claim, for they are human artifacts but they have their own biological interests.

However, we should note that the inherent worth possessed by synthetic life-forms presumably has no more impact on us than the inherent worth of a bacterium. This inherent worth is real; it exists. But it is not worth very much. Most of us have no qualms about destroying bacteria with mouthwash, so the inherent worth of synthetic life-forms would presumably have even less weight. Thus, the inherent worth of synthetic life-forms is unlikely to have significant ethical or social repercussions.

One final speculative remark deserves brief mention. If the right objective systemic properties really do capture the conditions for having biological interests or needs or purposes or the like, then the experimental program of synthetic biology can also be the basis for a constructive and creative research project in the origin and emergence of biological interests and other fundamental ethical concepts. We can create a variety of synthetic life-forms that span a broad spectrum of chemical complexity, and then we can note which kinds of chemical systems have which kinds of biological interests, needs, purposes, etc. This should enable us to determine exactly which kinds of chemical properties are required for the emergence of inherent worth. By this kind of program in "experimental ethics," top-down and bottom-up synthetic biology both enable us to direct investigative effort to answering questions about when certain values arise and norms emerge. This provides a new kind of empirical content for the exploration of deep and long-standing philosophical questions.

Conclusion

Environmentalism encourages an inclusive perspective about intrinsic values, in which human values and interests are only one part of the picture. The larger whole includes the inherent worth of other kinds of creatures, and the subjective and objective value of a variety of specific, systemic properties. We suggest that this perspective should be extended to apply to synthetic life-forms when they are examined with respect to intrinsic subjective value, intrinsic objective value, and inherent worth.

Certain kinds of intrinsic values drive arguments *against* creating synthetic life-forms. Creating them could cause cultural shock waves that undermine or conflict with important intrinsic values of certain cultures. Synthetic life-forms today also lack certain kinds of intrinsic objective values that natural life-forms possess (e.g., being wild or having a natural evolutionary history), and no synthetic life-form today can experience pleasure and pain.

Nevertheless, we conclude that synthetic life-forms still can possess each of the three kinds of intrinsic values. They can certainly be valued for their own sake, and so have (or could have) intrinsic subjective value for some people. Furthermore, synthetic life-forms have (or could have) many if not all of the objective, systemic properties that are thought, in environmental ethics, to provide intrinsic objective value and make life-forms intrinsically valuable. And synthetic life-forms have the basic biological interests, goals, and purposes that many people find sufficient grounds for inherent worth. So environmental ethics argues for the plausibility of applying the three kinds of intrinsic value to synthetic life-forms.

Here an important qualification arises. The same three kinds of intrinsic value are presumably possessed by *all* forms of life, including insects and bacteria, and most people have no compunction about doing whatever they want with insects and bacteria. So, presumably those people would have no compunction about doing whatever they want with any synthetic life-forms they might encounter.[10] The interests of synthetic life-forms would be as minuscule as the interests of bacteria and insects, and would be swamped when pitted against human interests.

Perhaps the strongest argument against our inclusive ethical perspective is that synthetic life-forms do not and could never have an important kind of intrinsic objective value of natural organisms. Note that this argument *against* inclusion in ethics would be championed by many of those who argue *for* inclusiveness in environmental ethics. We think this argument has at best contingent contemporary validity. While existing synthetic life-forms do not have the same kind of "wildness" or evolutionary history as natural living things, the difference might diminish and dissolve away entirely in the future. To see this, consider the following hypothetical scenario: Assume that many kinds of synthetic life-forms have been released in the environment a long time ago, and assume that the descendants of the synthetic life-forms intermingled and evolved along with the rest of the biotic community. The synthetic life-forms in this scenario become autonomous "living technology" (Bedau, Guldborg Hansen, et al. 2010; Bedau, McCaskill, et al. 2010), and populations evolve by natural selection. The descendants of synthetic life-forms have a complex and contingent evolutionary history, quite like the complex and contingent evolutionary histories of natural life-forms. So, a synthetic life-form could evolve into something with the intrinsic value that Leopold and his followers recognize in the natural environment and the naturally evolved life-forms in it.[11]

The environmental ethics literature often contrasts technological artifacts with natural organisms and biological communities, and assigns special value to the natural and biological. Synthetic biology proves that life-forms can also be artifacts designed and constructed through intentional human effort. This means that the living can no longer be equated with the natural. If the special value that environmentalists attribute to natural organisms comes from the fact that they are alive, then that same value is possessed by synthetic life-forms, even though they are synthetic, and even though they are artifacts.

Acknowledgments

For helpful comments, we thank Gregory Kaebnick, Emily Parke, Christopher Preston, Constance Putnam, Mark Sagoff, and Ron Sandler. The work reported here was supported by a Reed College Ruby Grant for Research in the Humanities.

Notes

1. But some of the properties thought to be pivotal for intrinsic value fail to fit synthetic life; an exceptionally good example is the property of having a natural evolutionary history, as we will see.
2. Preston reexamines whether synthetic life-forms have intrinsic value in chapter 6 of this volume.
3. Actually, Premise 2 is more complex, for there are situations in which artificial life-forms would have an evolutionary history, after they have been created and released into our environment and left alone to adapt and evolve just as all natural life-forms do. Here we ignore this possibility for the sake of simplicity.
4. We should note that Preston himself does not try to defend Premise 1; instead, he merely notes that it plays a strong role in the environmental ethics literature.
5. The general form of the conditional fallacy of denying the antecedent is this: If A then B; not A; so, not B.
6. These three kinds of intrinsic value are not mutually exclusive. To see this, consider that some objective, systemic analysis of biological autonomy and resilience is true. Now, it is common in environmental ethics to consider biological autonomy and resilience to have intrinsic objective value (see table 4.1). Furthermore, any system with biological autonomy and resilience will thereby also have interests of its own. Those interests include resiliently preserving oneself and maintaining one's autonomy. The system has these interests even if the system is neither conscious nor even sentient. Some in environmental ethics would say that valuers like us ought to care at least a bit, even if only a bit, about such interests in self-preservation and autonomy. From this perspective, a system that has interests of its own also has inherent worth, if only a slightly. And of course, systems whose objective systemic properties make them autonomous and resilient can certainly be subjectively valued for their own sake by people so inclined. So, one and the same systemic property could (make something) have intrinsic subjective value, intrinsic objective value, and inherent worth.

7. Preston (personal communication) has pointed out that Sandler's scheme is extremely inclusive, some might think to a fault, because it allows certain intrinsic values to be relational. On Sandler's scheme, if anyone happens to view or treat something with the right attitude, then it possesses intrinsic subjective value, and thus counts as having intrinsic value. This kind of intrinsic value seems to devolve into nothing more than personal preference. Do we want a scheme that includes intrinsic values that reflect nothing more than the preferences of psychopaths? There are various ways to reply to these questions by adjusting and qualifying Sandler's scheme. Exploring those issues is outside the scope of this chapter.

8. An explicit goal of much of bottom-up synthetic biology is to make protocells that possess and can sustain an integrated and mutually supporting triad of chemical processes, specifically the processes of programmed control, metabolism, and compartmentation (Rasmussen et al. 2004; Rasmussen, Bedau, McCaskill, and Packard 2009). Such mutually supporting chemical triads exhibit many of the objective systemic properties characteristic of minimal chemical life, such as the properties of autonomy, self-regulation, and evolution (Bedau 2010).

9. We are here ignoring the question of whether we should also consider the interests of the organism's kin and of its whole lineage.

10. Of course, manipulating and destroying whole insect and bacterial species could threaten the well-being of humans, and so would deserve moral scrutiny.

11. Gregory Kaebnick (personal communication) has suggested two further considerations that support our contention that synthetic organisms can become "natural" in the relevant sense. It is sometimes suggested that life on Earth originated elsewhere in the universe, and was seeded on Earth through a meteorite. Consider a hypothetical case in which all life on Earth was planted here deliberately by some earlier civilization. In this case, all life on Earth is an artifact, or at least originated as an artifact. In this hypothetical case, all life on Earth ends up having a complex and contingent evolutionary history. Furthermore, that evolutionary history is exactly the same as if all life on Earth arose here spontaneously. This shows that artifacts (or at least their descendants) can become natural in any sense that is relevant to environmental ethics. Now consider a second hypothetical case. We are already genetically engineering American chestnuts to have the blight resistance of Asian chestnuts. Suppose that we synthesize the entire American chestnut genome with the Asian chestnut genes and grow trees with that genome as an effort to restore some American chestnut woods. Those trees would fit right into an existing ecological niche, and they would become a natural part of the environment.

References

Andrianantoandro, E., S. Basu, D. K. Karig, and R. Weiss. 2006. Synthetic biology: New engineering rules for an emerging discipline. *Molecular Systems Biology* 2: 1–14.

Attfield, R. 1981. The good of trees. *Journal of Value Inquiry* 15:35–54.

Bedau, M. A. 2010. An Aristotelian account of minimal chemical life. *Astrobiology* 10:1–10.

Bedau, M. A., G. Church, S. Rasmussen, A. Caplan, S. Benner, M. Fussenegger, J. Collins, and D. Deamer. 2010. Life after the synthetic cell. *Nature* 465:422–424.

Bedau, M. A., Pelle Guldborg Hansen, Emily Parke, and Steen Rasmussen (eds.). 2010. *Living Technology: 5 Questions*. Copenhagen: Automatic Press/VIP.

Bedau, M. A., J. S. McCaskill, N. H. Packard, and S. Rasmussen. 2010. Living technology: Exploiting life's principles in technology. *Artificial Life* 16:89–97.

Bedau, M. A., and E. C. Parke (eds.). 2009. *The Ethics of Protocells: Moral and Social Implications of Creating Life in the Laboratory*. Cambridge: MIT Press.

Bedau, M. A., E. C. Parke, U. Tangen, and B. Hantsche-Tangen. 2009. Social and ethical checkpoints for bottom-up synthetic biology, or protocells. *Systems and Synthetic Biology* 3:65–75.

Boldt, J., and O. Müller. 2008. Newtons of the leaves of grass. *Nature Biotechnology* 26:387–389.

Bookchin, M. 1988. Social ecology versus deep ecology. *Socialist Review* 88:11–29.

Callicott, B. 1989. *In Defense of the Land Ethic: Essays in Environmental Philosophy*. Albany: State University of New York Press.

Carlson, Robert H. 2010. *Biology Is Technology: The Promise, Peril, and New Business of Engineering Life*. Cambridge: Harvard University Press.

Caschera, F., G. Gazzola, M. A. Bedau, C. Bosch Moreno, A. Buchanan, J. Cawse, N. Packard, and M. M. Hanczyc. 2010. Automated discovery of novel drug formulations using predictive iterated high throughput experimentation. *PLoS ONE* 5 (1):e8546.

Cho, M. K., D. Magnus, A. L. Caplan, D. McGee, and the Ethics of Genomics Group. 1999. Ethical considerations in synthesizing a minimal genome. *Science* 286:2087–2090.

Devall, B., and G. Sessions. 1985. *Deep Ecology: Living as if Nature Mattered*. Layton, UT: Gibbs Smith.

Elliot, R. 1992. Intrinsic value, environmental obligation and naturalness. *Monist* 75:138–160.

Endy, D. 2005. Foundations for engineering biology. *Nature* 438:449–453.

Feinberg, J. 1974. The rights of animals and unborn generations. In *Philosophy and Environmental Crisis*, ed. W. Blackstone. Athens: University of Georgia Press.

Garfinkel, M. S., D. Endy, G. L. Epstein, and R. M. Friedman. 2007. Synthetic genomics: Options for governance. Final report of project funded by the Sloan Foundation. http://www.jcvi.org/cms/fileadmin/site/research/projects/synthetic-genomics-report/synthetic-genomics-report.pdf.

Gibson, D. G., G. A. Benders, C. Andrews-Pfankoch, E. A. Denisova, H. Baden-Tillson, J. Zaveri, T. B. Stockwell, et al. 2008. Complete chemical synthesis, assembly, and cloning of a *Mycoplasma genitalium* genome. *Science* 319:1215–1220.

Gibson, D. G., J. I. Glass, C. Lartigue, V. N. Noskov, R.-Y. Chuang, M. A. Algire, G. A. Benders, et al. 2010. Creation of a bacterial cell controlled by a chemically synthesized genome. *Science* 329:52–56.

Gould, S. J. 1992. *The Panda's Thumb: More Reflections in Natural History*. New York: Norton.

Hargrove, Eugene C. 1989. *Foundations of Environmental Ethics*. Englewood Cliffs, NJ: Prentice-Hall.

Hettinger, N., and B. Throop. 1999. Refocusing ecocentrism: De-emphasizing stability and defending wildness. *Environmental Ethics* 21:3–21.

Kaebnick, G. E. 2009. Should moral objections to synthetic biology affect public policy? *Nature Biotechnology* 27:1106–1108.

Katz, E. 1992. The call of the wild. *Environmental Ethics* 14:265–273.

Krieger, M. 1973. What's wrong with plastic trees? *Science* 179:446–455.

Lartigue, C., J. I. Glass, N. Alperovich, R. Peiper, P. P. Parmar, C. A. Hutchinson III, H. O. Smith, and J. C. Venter. 2007. Genome transplantation in bacteria: Changing one species into another. *Science* 317:632–638.

Leopold, A. 1949. *A Sand County Almanac*. New York: Oxford University Press.

Light, A. 2000. Ecological restoration and the culture of nature. In *Restoring Nature: Perspectives from the Social Sciences and Humanities*, ed. P. Gobster and B. Hall. Washington, DC: Island Press.

Midgley, M. 1983. Duties concerning islands. *Encounter* 60:36–43.

Nolt, J. 2009. The move from *is* to *good* in environmental ethics. *Environmental Ethics* 31:135–154.

Norton, B. 1991. *Toward Unity among Environmentalists*. New York: Oxford University Press.

O'Neill, J. 1993. *Ecology, Policy and Politics: Human Well-Being and the Natural World*. New York: Routledge.

Presidential Commission for the Study of Bioethical Issues. 2010. *New Directions: The Ethics of Synthetic Biology and Emerging Technologies*. http://www.bioethics.gov/documents/synthetic-biology/PCSBI-Synthetic-Biology-Report-12.16.10.pdf.

Preston, C. 2008. Synthetic biology: Drawing a line in Darwin's sand. *Environmental Values* 17:23–39.

Rasmussen, S., M. A. Bedau, L. Chen, D. Deamer, D. C. Krakauer, N. H. Packard, and P. F. Stadler. 2009. *Protocells: Bridging Nonliving and Living Matter*. Cambridge, MA: MIT Press.

Rasmussen, S., M. A. Bedau, J. M. McCaskill, and N. H. Packard. 2009. Roadmap to protocells. In *Protocells: Bridging Nonlivng and Living Matter*, ed. S. Rasmussen, M. A. Bedau, L. Chen, D. Deamer, D. C. Krakauer, N. H. Packard, and P. F. Stadler, 71–100. Cambridge, MA: MIT Press.

Rasmussen, S., L. Chen, D. Deamer, D. C. Krakauer, N. H. Packard, P. F. Stadler, and M. A. Bedau. 2004. Transitions from nonliving to living matter. *Science* 303:963–965.

Ro, D.-K., E. M. Paradise, M. Ouellet, K. J. Fisher, K. L. Newman, J. M. Ndungu, K. A. Ho, et al. 2006. Production of the antimalarial drug precursor artemisinic acid in engineered yeast. *Nature* 440:940–943.

Rolston, H., III. 1982. Are values in nature subjective or objective? *Environmental Ethics* 4:125–151.

Rolston, H., III. 1988. *Environmental Ethics: Duties to and Values in the Natural World*. Philadelphia: Temple University Press.

Russow, L. 1981. Why do species matter? *Environmental Ethics* 3:101–112.

Sagoff, M. 1997. Do we consume too much? *Atlantic Monthly* , June, 80–96.

Sandler, R. 2007. *Character and Environment: A Virtue-Oriented Approach to Environmental Ethics*. New York: Columbia University Press.

Sandler, R. 2010. Is artefactualness a value-relevant property of living things? *Synthese*.

Schmidt, M. 2009. Societal aspects of synthetic biology. Special issue, *Systems and Synthetic Biology* 3:1–2.

Schmidt, M., A. Kelle, A. Ganguli-Mitra, and H. de Vriend. 2010 (eds.). *Synthetic Biology: The Technoscience and Its Societal Consequences*. Dordrecht: Springer.

Schmidtz, D., and E. Willott. 2002. *Environmental Ethics: What Really Matters, What Really Works*. New York: Oxford University Press.

Singer, Peter. 1975. *Animal Liberation*. New York: Harper.

Sober, E. 1986. Philosophical problems for environmentalism. In *The Preservation of Species: The Value of Biological Diversity*, ed. B. Norton, 173–194. Princeton: Princeton University Press.

Specter, M. 2009. A life of its own: Where will synthetic biology lead us? *New Yorker*, September 28, 56–65.

Stone, C. 1972. *Should Trees Have Moral Standing?* Los Altos, CA: William Kaufman.

Szostak, J. W., D. P. Bartel, and P. L. Luisi. 2001. Synthesizing life. *Nature* 409:387–390.

Taylor, P. 1981. The ethics of respect for nature. *Environmental Ethics* 3:197–218.

Taylor, P. 1986. *Respect for Nature: A Theory of Environmental Ethics*. Princeton: Princeton University Press.

Van Est, R., H. de Vriend, and B. Walhout. 2007. *Constructing Life: The World of Synthetic Biology*. The Hague: Rathenau Instituut.

Varner, G. 2002. Biocentric individualism. In *Environmental Ethics: What Really Matters, What Really Works*, ed. D. Schmidtz and E. Willott, 108–119. New York: Oxford University Press.

Yokobayashi, Y., R. Weiss, and F. H. Arnold. 2002. Directed evolution of a genetic circuit. *Proceedings of the National Academy of Sciences of the United States of America* 99:16587–16591.

5

Three Puzzles Regarding the Moral Status of Synthetic Organisms

John Basl and Ronald Sandler

An artifact is an entity designed and created by an intentional agent who is capable of imagining the world as it might otherwise be, evaluating the alternatives, and devising strategies for realizing them.[1] A living thing is artifactual, not natural, to the extent that it is designed and engineered by an agent. The difference between artifactual organisms and natural organisms is thus in their origins, and it is a matter of degree.

Minimally artifactual organisms are commonplace. Selective breeding, grafting, and intentional hybridization—processes that have been occurring since the beginning of agriculture—produce minimally artifactual organisms. But although traditional techniques have been immensely successful in producing organisms with desirable and useful traits, they have significant "engineering" limitations. There is a lack of control over which traits offspring receive from each parent, there are constraints on the possible genetic recombinations, and there is a limited range of deviation from the base life-forms.

Beginning with the development of recombinant DNA techniques in the 1970s, these constraints have been increasingly loosened. Recombinant DNA techniques enable the isolation of genes that code for desired traits in individuals of one species and the insertion of those genes into the genome of another species. In this way, genetic material from individuals of one species can be inserted into the genome of individuals of another species to create organisms with genomic material that could not have combined in the absence of intentional gene-level intervention—for example, chimpanzees with jellyfish genes, goats with golden orb spider genes, rice with maize genes, maize with bacteria genes, mice with human genes, and salmon with ocean pout genes.

The knowledge base and technology needed for genomic sequencing, isolating genes, determining gene functions, knocking out genes, and assembling genomic material, while still quite imperfect, has progressed to the point that it is possible to intensively engineer genomes using elements from multiple biological (and potentially artificial) genetic sources. For example, one research group has engineered an *E. coli* that produces over 10,000 times more artemisinic acid—the precursor for artemisinin, an antimalarial drug—than a natural bacteria. They accomplished this by transplanting genes from the wormwood plant (*Artemisia annua),* the traditional source for artemisinin, and yeast (*Saccharomyces cerevisiae*), which codes for the requisite metabolic processes, into the bacteria, as well as by disabling genes in the bacteria that code for interfering processes. The result is an organism in which the yeast's metabolic process utilizes the wormwood information to produce a desired product in an engineered bacterial host (Ro et al. 2006). Thus, the trend in genomic design and construction—what is known as synthetic biology or synthetic genomics—is that the base organism is less and less a constraint on what can be created. In fact, the ideal base organism for synthetic biology is one that has been stripped down to a "chassis": all genetic material and metabolic processes not essential to survival, maintenance, and reproduction have been removed, and the desired processes can be inserted. The goal of some researchers is that there be a vast, ever-expanding repository of biological (and nonbiological) parts and assembly instructions that can be used to efficiently, precisely, and (relatively) easily design and construct organisms with desired functionalities. Thus, synthetic biology, unlike hybridization or even recombinant DNA techniques alone, can produce thoroughly engineered, highly artifactual synthetic organisms.[2]

If that happens, then some interesting new puzzles will arise having to do with the moral status of synthetic organisms. It is a common view among environmental ethicists and bioethicists that at least some nonsentient living things—such as plants, trees, and people in permanent vegetative states—have a good of their own or interests that moral agents need to take into consideration (Goodpaster 1978; Taylor 1989; Varner 1998; Sandler 2007). That is, they can be directly benefited or harmed, and they are therefore not mere *things*, with only instrumental or derivative value. Instead, they matter in themselves, for what they are. We will hereafter refer to this type of value as *inherent worth*. (As we are using the term, *inherent worth* is a variety of *intrinsic value*. Something has intrinsic value if it is valuable for what it is and as an end, but it has inherent worth if and only if its interests need to be considered, for its own sake, in deliberations regarding actions, practices, and policies that might affect it.[3])

It is also a common view that artifacts cannot be directly benefited or harmed and have only instrumental or derivative value; they do not have inherent worth. What, then, are we to make of synthetic organisms? Since they are living, they seem to be candidates for inherent worth, but since they are also highly artifactual, they do not seem to be candidates for inherent worth. Do they possess inherent worth or not?

Answering these questions requires becoming clear on the basis for the inherent worth of nonsentient living things. In the next two sections of the chapter, we argue that the only reasonable account of how nonsentient living things have a good of their own is an *etiological* account, in which their good is grounded in what their parts and processes were selected for. Like natural organisms, synthetic organisms have a selection history, just one that is, at least in part, artificial rather than natural. Does this affect whether they have a good of their own or, if they have a good of their own, whether they have inherent worth? We argue that, given the etiological account of functions, it should not make a difference.

But then several puzzles arise regarding the moral status of synthetic organisms, nonliving artifacts, and natural organisms, and we consider these puzzles in the three sections that follow. The first puzzle is that entities that look to be identical could differ in their moral status: one could have inherent worth, while the other lacks it. The second puzzle is that artifacts appear to have a good of their own, contrary to the common view that artifacts have only instrumental value. The third puzzle (which builds on the second) is that artifacts appear to have inherent worth. We conclude with a discussion of the significance of these puzzles and the practical relevance of the moral status of synthetic organisms.

The Good of Nonsentient Organisms

Determining which entities have inherent worth or are directly morally considerable—whether only humans, only sentient animals, all living things, or environmental collectives (such as species and ecosystems)—is a prominent issue in environmental ethics. To attribute inherent worth to an entity is to say that (a) it has a good of its own, and (b) its good should be taken into account in moral deliberations for its own sake. To judge that, for example, primates have inherent worth is to judge that they have goods of their own—they can be benefited or harmed—and that we must be sensitive to their good, independent of our own interests or how we feel about them, in deliberations regarding actions, practices, and policy that could affect them.

Peter Singer (1989) has famously claimed that all and only sentient beings have inherent worth. Singer argues that all sentient beings, like most human beings, have goods grounded in their psychological capacities. The range of interests an entity has depends on its range of capacities, but, Singer claims, all sentient beings have at least an interest in pleasure and the avoidance of pain. Singer then argues that there is no morally relevant difference between human and nonhuman sentient individuals that would justify dismissing or discounting their like interests.

Singer denies that nonsentient individuals have inherent worth on the grounds that, lacking cognitive capacities, they do not have goods of their own. Singer's argument isn't that nonsentient individuals have a good, but that it is irrelevant to our moral deliberations. Rather, he thinks that there is no sense in which nonsentient entities can be benefited or harmed at all. On his view, they do not have a welfare.

Welfare is often understood as an element of well-being that is tied, at least partially, to the mentalistic components of an individual's life. An individual has a high welfare to the extent that it has a high ratio of positive mental states to negative mental states (Feldman 2004), or to the extent that its desires (or informed desires) are satisfied (Streiffer and Basl 2011). If welfare is understood in this fashion, it is true that nonsentient individuals have no welfare. However, it is controversial whether welfare is the sole component of well-being, and many philosophers prefer "objective list" views of human or animal well-being (Hursthouse 1999; Sandler 2007; Streiffer and Basl 2011; Griffin 1988). On such views, pleasure, pain, suffering, enjoyment, or the satisfaction of desires are components of well-being but are not the whole story. For example, many people think that a life goes better, all else equal, if it is authentic or if an individual's experiences are veridical; it is better, for example, not just to think one has good friends and is respected, but also actually to have good friends and be respected (Nozick 1974). And others believe that nonsubjective health is a component of well-being (Taylor 1989; Varner 1998; Hursthouse 1999; Sandler 2007).

Insofar as objective list views are plausible, Singer's inference from an entity's lacking a welfare to its having no good of its own, and therefore lacking inherent worth, is invalid. It is true that nonsentient individuals take no interest in things; they do not care how their life goes, and they recognize nothing as contributing to their good. However, it does not follow that nothing is in their interests (Taylor 1989; Sandler 2007). Moreover, there is an intuitive case for the claim that living things have a good of their own—that they can be benefited or harmed in ways that do not require appealing to the good or interests of other entities. For example, if the roots of an oak tree are ripped up during an excavation, it is, in a straightforward way, bad for and harmful to the oak tree. It impairs its parts and processes, and thereby its capacities to survive, grow, and reproduce. Moreover, it is possible to make sense of the harm without having to refer to anything beyond the oak tree, such as the owner of the tree or the services that it provides for people or other species. It is bad for the oak tree because it diminishes *its* capacity to pursue *its* ends (or to flourish as an oak tree), independent of the effects on other organisms. And there is nothing special about oak trees in this respect. All natural living things can be benefited (or harmed) in this sense. Devil facial tumor disease is bad for Tasmanian devils, oak wilt is bad for oak trees, and amoxicillin is bad for streptococci, independent of what they do for us (or for individuals of any other species) or how we feel about them.

Nevertheless, for the intuitive argument presented above to be complete, and the view that nonsentient entities have inherent worth to be substantiated, there must be a nonarbitrary, nonderivative way to ground the good of living beings. The most prominent approach to doing so is to ground the good of nonsentient entities in their teleological organization (Goodpaster 1978; Taylor 1989; Varner 1998; Sandler and Simons 2012). Living organisms, sentient and nonsentient, are goal-oriented systems with parts and processes that have the purpose of realizing certain ends for the organism as a whole. Once an entity is understood as goal-directed, the content of its good can be understood in terms of what promotes its ends. For example, one of the ends of the leaves of plants is to gather energy from sunlight; therefore, it is good for a plant that it have access to sunlight.

The Etiological Account of Teleological Organization

Appeals to teleological organization are a step toward explicating the basis and content of nonsentient organisms' good. However, it is also necessary to provide an explanation for the teleological organization; one that demonstrates that the teleological organization of plants isn't merely imagined. The teleological organization of sentient beings can be grounded in their psychology. However, the teleological organization of nonsentient individuals must be grounded in some other way. The most prominent and most plausible explanation of the source of teleology in nonsentient biological organisms appeals to an etiological account of functions (Cahen 2002; Varner 1998).

According to etiological accounts of function, a part or trait of an organism has the function of doing F only if it was selected for doing F.

For example, the function of the heart is to pump blood only if there was selection for pumping blood. In the case of nonartifactual organisms, the relevant selection process is natural selection. One of the primary motivations in the development of etiological accounts of function is that they seem the best way to make sense of certain teleological attributions about the nature of organisms, their parts, and their processes (Wright 1994; Millikan 1989, 1999; Neander 1991, 2008; Mitchell 1993). Alternative accounts of function, such as goal-contribution accounts (Boorse 1976) or causal-role accounts (Cummins 1975), explain functions in terms of what a part or process does in an organism or system. These accounts are not well suited to grounding teleology, because they either assume teleology or they merely describe what a part (or process) does. Etiological accounts, on the other hand, can ground teleology by appealing to the consequences or effects for which a part was selected. By doing so, etiological accounts can not only ground teleology but also distinguish the purpose of parts from mere by-products (Wright 1973).

That nonsentient organisms are teleologically organized, and thereby have a good of their own, does not settle the question of whether they have inherent worth. There is the further question of whether human beings should be responsive or sensitive to the goods of nonsentient individuals. The most prominent arguments in favor of the moral considerability of nonsentient organisms employ a strategy much like Singer's. The strategy is to first establish that nonsentient beings have interests and then argue that there is no morally relevant difference between nonsentient beings and sentient beings that would justify discounting like interests. Varner (1998) has argued that there are cases where we are required to be sensitive to the interests of humans and nonhuman animals, even when the interests cannot be captured in terms of (or reduced to) their mental states. He concludes that in those cases, there are morally relevant interests grounded in the biology of organisms—grounded, that is, in the etiological account of functions. Insofar as these interests are morally relevant, so should be similar interests in nonsentient organisms. To exclude them would be arbitrary and unmotivated. This is not to claim that the interests of nonsentient organisms are the same as those of sentient organisms or that the interests of nonsentient organisms cannot be traded off against the interests of sentient ones. Sentient beings have a range of interests that nonsentient beings lack. Being responsive to all the interests of both sentient and nonsentient organisms will often mean sacrificing the interests of nonsentient organisms for the good of sentient ones.

In what follows we discuss a series of puzzles, relevant to the inherent worth of nonsentient synthetic organisms, that arise from etiologically grounding claims about the good of nonsentient organisms.

Puzzle One: The Etiological Account of Teleology and Intrinsically Identical Beings

The etiological account of teleology is the best available account of the goods of nonsentient beings, but it raises several puzzles that become manifest when one attends to artifactual organisms. One puzzle is what we might call the *symmetry puzzle*. It seems reasonable to expect that two intrinsically identical entities—entities that have the same internal properties—should have the same good. If one has a good of its own, so should the other, and if they have a good of their own, the content of their good should be the same as well. If a particular rose has a good of its own, it seems reasonable to think that an identical rose will also have a good of its own (and that the content of the good will be similar), and that if the first rose lacks a good of its own, so, too, will its twin. It also seems plausible that the symmetry would apply to inherent worth—that two internally identical entities should either both have or both lack inherent worth.

However, given the etiological account of teleology, the symmetry theses described above are false. According to the etiological account of teleology, the good of a plant is a function of what it, its parts, and its processes were selected for. Selection processes need not be preserved across intrinsic duplicates. Consider an artifactual rose, one that is the product of synthetic biology, but internally identical to a naturally occurring rose. If its parts and processes were selected for reasons other than promoting the growth, survival, and reproduction of the organism (which are, at a broad level of description, the bases on which the parts and processes of naturally occurring organisms are selected), then the good of the synthetic rose will diverge from that of the natural rose.[4] This result is possible because, on the etiological account of interests, an entity's good depends on relational properties—its selection history.

Given the etiological account of interests, it is even possible that an entity has a good of its own while its intrinsically identical duplicate lacks one altogether. Consider an instant organism, one that comes to exist by completely random chance, as if a whirlwind had whipped together just the right chemicals and materials to produce something intrinsically identical to a rose. According to the etiological account of teleology, the instant organism will lack a good of its own. Because its parts and processes have no selection history, it has no goals or ends toward which it strives (no teleological organization), and therefore nothing is good or bad for it. This is despite the fact that there are things that are good or bad for the naturally evolved rose, which is identical to the instant rose in terms of all its intrinsic properties. Furthermore, if the real rose has inherent worth, then the inherent worth version of the symmetry thesis is also false. Because the instant rose lacks a good of its own, it is not a candidate for having inherent worth.

The etiological account of teleology is the best available account of the good of nonsentient organisms. Furthermore, it seems intuitive that things are good or bad for traditional nonsentient organisms. In order to preserve these intuitions, it is necessary to deny the symmetry theses, even though they are also intuitively plausible. This is the first puzzle.

Puzzle Two: The Good of Artifacts

The etiological account of teleology requires only that an entity be teleologically organized in order to have a good of its own. This produces a puzzle concerning artifacts. While many people think that organisms, even nonsentient ones, have a good of their own, few think that artifacts have a good of their own. Yet artifacts are teleologically organized according to the etiological account of teleology. Artifacts, and their parts and processes, are goal-directed; they have been selected for particular ends. A car has the parts and processes that it does, organized as they are, because of the role they play in locomotion. The wheels, axles, engine, cooling system, and transmission were selected for because they are conducive to bringing about that end. Thus, on the etiological account of how the good of nonsentient things are generated, a car has a good of *its own*. This is contrary to the widely shared view that artifacts are only good or bad derivatively or as means to an end. Thus, it seems, one must either reject the etiological account, and thereby give up the view that plants and trees have a good of their own, whether they are natural or synthetic, or one must accept that nonliving artifacts, such as candy wrappers, picture frames, and iPods, have a good of their own. In the latter case, of course, synthetic organisms will have a good of their own as well.

This puzzle has not gone unnoticed by those who argue for grounding the goods of nonsentient organisms in etiological accounts of teleology. In response, most accept that artifacts are teleologically organized but deny that teleological organization is sufficient for an entity's having a good of its own (Taylor 1989; Varner 1998). They claim that there is some difference between nonliving artifacts and organisms such that only the latter have a good of their own. The differences most commonly appealed to, however, are wanting.

One difference, of course, is that only organisms are living. Does the living/nonliving distinction explain why organisms have a good of their own and artifacts do not? Despite some intuitive appeal, it does not. The issue is that the etiological account of teleology seems to have the counterintuitive implication that soda cans and cell phones have a good of their own, in the same sense as nonsentient organisms. The claim that being alive is a relevant difference solves the puzzle in that it avoids the problematic implication. However, if no reasons as to why it is a relevant difference are given, then it is completely ad hoc and question-begging. It just asserts that the difference at issue, which appears not to be relevant, is relevant.

Is there any justification for thinking that the living/nonliving distinction is relevant, and so not just an ad hoc, question-begging response? Such a justification would have to appeal to some feature of living things that artifacts lack. But the most promising internal features—namely, that organisms are internally organized and goal-directed—have already been considered. The problem is that artifacts have these properties to the same extent as nonsentient organisms (natural and synthetic).

Another feature of organisms is that they are dynamic systems. They are responsive to perturbations, they attempt to self-repair, and they metabolize. Synthetic organisms are similarly dynamically organized. Is an entity's being dynamic a plausible supplementary criterion to teleological organization for an entity to have a good of its own, one that would explain why all nonsentient living things have a good of their own, even if they are artifactual? It is not. Many traditional artifacts are dynamic systems. Computers (being artifacts) do not metabolize, but they depend on external resources (such as electricity) to operate; they respond to changes in their condition in order to maintain a certain range of states (for example, the processor fan speeds up or slows down in response to changes in internal temperature); and they attempt to preserve and repair themselves when they are exposed to potential threats or when damages occur (for example, detecting and removing potential viruses, quarantining contaminated files, and preventing hardware failures). The requirement of dynamic organization will rule out some artifacts from having a good of their own, but many others—cell phones, computers, cars, and thermostats, among them—will still satisfy the condition. (We discuss in the next section the relationship between dynamism and inherent worth.)

Perhaps the most common way of distinguishing artifacts from nonsentient organisms is to claim that the ends of artifacts are derivative in a way that the ends of organisms are not. After all, artifacts are created by humans for our use, while organisms evolved independently of us. Furthermore, the goal-directedness of artifacts cannot be explained except by reference to the intentions and ends of human beings. Therefore, it seems that the ends or goals of artifacts are not their own, but ours. If this is right, then the puzzle of artifacts having a good of their own is resolved by their derivative nature.

The ends of artifacts are claimed to be derivative in two distinct ways. One is in terms of their use. Artifacts are created expressly to be used by humans to achieve human ends. Most organisms evolved independently of our wishing to use them for our own ends. However, many organisms exist solely because they help us achieve our ends. Organisms used in agriculture, in science, in recreation, and as companions are in this category. Of course, these organisms share an evolutionary history with organisms not engineered by us. In the future, however, we may create similar organisms from scratch. For these reasons, the distinction between being created for use and not being created for use does not track the distinction between artifacts and organisms. Therefore, it cannot resolve the puzzle without also denying that some living things, including synthetic organisms, have a good of their own.

The other way that the ends or goals of artifacts are derivative is that any explanation of the ends or goals of artifacts references the intentions of conscious agents. This response to the puzzle acknowledges that the parts of both organisms and artifacts are selected for certain ends, but claims that because one is the result of natural selection whereas the other is selected by conscious agents, organisms have a good of their own while artifacts do not. One implication of this response, it should be noted, is that, if it works, *synthetic* organisms do not have a good of their own (at least in the first generations).

The problem with this line of reasoning is that to distinguish artifacts from organisms on the basis of a difference in how their teleological organization is explained conflates the *explanation* of teleological organization with the *subject* of teleological organization. The explanations of many of our own ends reference the intentions of others. Human ends are influenced by our educational backgrounds, opportunities, and resources, for example, which are often the result of others' intentions. The explanation for why Tiger Woods became a golfer requires that we reference his father's intentions. Still, his golf-oriented ends are part of his good (and teleological orientation), distinct from his father's. Our ends, whether psychological or biological, are our own, independent of the explanation for how they came to be our ends. Similarly, artifacts are themselves teleologically organized, and their ends are their own, regardless of whether our intentions are part of the explanation for how they came to be that way. Thus, pointing out that artifacts are the product of human intentions or are created to be used by us does not justify denying that they have a good of their own. Therefore, derivativeness does not resolve the puzzle; it does not justify attributing a good to nonsentient natural living organisms but not to nonliving (or living) artifacts. This is the second puzzle.

Each of the differences between artifacts and nonsentient organisms offered as a reason for denying that artifacts have a good of their own fails to do the requisite work. Either the difference does not distinguish cleanly between artifacts and nonsentient organisms (and depending on the difference, synthetic organisms are sometimes grouped with artifacts, sometimes with organisms) or else the difference is not an adequate basis for distinguishing between entities that have a good of their own and those that do not. The puzzle of artifacts therefore remains. It seems that the etiological account requires attributing a good to both nonsentient organisms (synthetic and natural) and artifacts. The alternative is to deny the etiological account, in which case neither nonsentient organisms nor artifacts have a good of their own. Artifacts and nonsentient organisms appear to be in it together.

Puzzle Three: The Moral Status of Artifacts

That artifacts have a good of their own may seem odd, but the mere fact that artifacts have a good of their own does not imply that we ought to care about their good—it does not imply, that is, that they have inherent worth (Taylor 1989; O'Neill 2003), let alone determine how we ought to respond to it (Sandler 2007). It would be odder still if we had to care about the good of soda cans, computers, and toasters for their own sake. Thus another puzzle arises: how can nonsentient living things with a good of their own have inherent worth, while nonliving artifacts with a good of their own do not?

If nonsentient natural living things have a good of their own that we ought to care about, then the view that we need not care about the good of nonsentient, artifactual living things or nonsentient, nonliving artifacts is justified only if there is a relevant difference between natural and artifactual, or between living and nonliving, that justifies caring about one and not the other. Another way to put this is to ask whether "artifactual" and "nonliving" are morally relevant properties with respect to inherent worth.[5]

It is possible to create an artifactual organism that is qualitatively identical to a natural organism (Gibson et al. 2010). If there is a basis for disregarding the good of one but not the other of the organisms, it would have to be a consideration extrinsic to the organisms—a consideration not related to internal properties). However, part of the concept of inherent worth is that it is not based on extrinsic properties. It is a type of value that has to do with the features of an entity itself—its complexity, capacities, and interests or good. It is not based on an entity's relational properties—its utility or why or how it originated. It is the status that an entity has in virtue of what it is. Because the artifactualness of an artifactual organism is an extrinsic property, it is not an appropriate basis for disregarding the good of an organism.[6]

The argument that artifactualness is not a morally relevant property with respect to inherent worth implies that if nonsentient natural living things have inherent worth, then nonsentient artifactual living things of comparable complexity or capacities have the same inherent worth. That is, synthetic bacteria have the same inherent worth as naturally occurring bacteria, and synthetic plants have the same inherent worth as naturally occurring and evolved plants.

This may not be as counterintuitive as it first appears. They are all living things, and it may be that plants and bacteria, even if they have inherent worth, do not merit much if any concern. Perhaps the responsiveness from moral agents that their worth justifies is insignificant. In fact, proponents of the inherent worth of natural living things typically argue that this is the case with respect to microorganisms. After all, it is not possible to be a human being in the world without relentlessly harming very large numbers of them. Our immune systems do that all the time. Moreover, it is not clear how one could even take their good into account, given their size, multitude, and diversity. We are, in practical terms, not really agents with respect to them. For this reason, even proponents of the view that all living things have inherent worth do not believe that we should (or even can) be responsive to the good of microorganisms (Sandler 2007; Kawall 2008).

As the case of microorganisms illustrates, the fact that an entity has inherent worth does not itself determine how much weight we ought to give its good or how we ought to respond to it. These depend as well on our form of life—our scale of agency, capabilities, and dependencies and vulnerabilities. For example, the fact that we must consume living things to survive (with some extreme exceptions) implies that respect for plants cannot require that we always refrain from killing or using them. It does not follow from this that all killing and using is respectful; some might be unnecessary and wanton. But it does indicate that responsiveness to inherent worth depends on facts about our form of life and what we are capable of. How we ought to respond to the inherent worth of other organisms also depends on facts about them. That compassion is appropriate to sentient beings with inherent worth does not imply that it is appropriate to plants, which lack the capacity to suffer. Therefore, if artifactual living things have inherent worth, it does not follow that they are due the same sorts of responsiveness and consideration (let alone treatment) as sentient beings or rational beings. So, perhaps, allowing that artifactualness is not a relevant property with respect to inherent worth is not so puzzling after all.

What of the relevance of "living" to inherent worth? The mere fact that something is nonliving appears to be an ad hoc reason for discounting its good. One can reasonably ask, why should being alive make a difference? One possible reason is that living things are metabolic or dynamic in a way that, for example, soccer balls and lampposts are not. However, as discussed earlier, there are artifacts that are also "metabolic" in the sense of being dynamic (and not merely static), and it is not clear what the metabolic nature of living things captures that might be morally relevant if not the dynamic and processing aspects. Thus, if "dynamism" is required for an entity's good to be morally considerable, then some artifacts will be ruled out, but many others will not be. Moreover, not all living things are metabolically equal. Compare bristlecone pines, which have very low metabolic rates, with hummingbirds, or even bamboo. If a minimal level of metabolism (or dynamism) is necessary for an entity with a good of its own to be regarded as possessing inherent worth, then the lower the minimum, the more inclusive the standard will be with respect to living things (natural and artificial) as well as nonliving things.

There is also the difficulty of explaining why metabolic activity or dynamism ought to be regarded as a necessary condition for inherent worth. If nonmetabolic (or nondynamic) entities have a good of their own, why should their good not be considered? Perhaps what the above discussion points toward is not a metabolic/dynamism baseline for having inherent worth, but that for entities that are not dynamic (and so are comparatively very low with respect to complexity and capacities), their good simply does not count much in comparison to that of nonsentient, dynamic living things (which in turn might not count much in comparison to sentient living things). This would imply that "living" is not a value-relevant property with respect to inherent worth, but that nonetheless, the worth of (some) nonliving things is highly insignificant. This would avoid the problem of our having to take the inherent worth of soccer balls, garage doors, and printers into account, but it would not eliminate the moral status puzzle. This is because nonliving, nondynamic artifacts would have inherent worth (which is itself odd), but their inherent worth would not matter for all (or nearly all) practical purposes. That is, there would be entities with inherent worth that do not really matter, which seems even more puzzling, since inherent worth is about what things matter in themselves.

The moral status puzzle concerning artifactual organisms and nonliving things arises when one asks whether there is any reason that the good of natural living things should be considered, but that the good of artifacts or nonliving things should not. To assert that the properties "nonliving" and "artifactual" are relevant to inherent worth is ad hoc and question-begging. Moreover, any plausible attempt to explicate the basis for their being relevant fails to cleanly divide natural living things on the one hand and nonnatural living things or nonliving things on the other hand. There seems to be no principled way to count all nonsentient living things (or even all plants) as having inherent worth while also excluding all nonliving things (or all nonnatural things). This is the third puzzle.

Conclusion

If nonsentient living things have inherent worth (as many environmental ethicists and bioethicists argue or presuppose), then they must have a good of their own. The only reasonable account of what grounds their good is etiological. On all other approaches, claims about the content of their good are arbitrary or derivative. However, the etiological account of what grounds an entity's good (or interests) is neutral with respect to the type of selection process that generates the goods. Therefore, if nonsentient natural organisms have a good of their own, then so, too, do synthetic organisms. However this gives rise to several puzzles: that internally identical organisms can have different goods (because they have different etiologies), that nonliving artifacts have a good of their own, and that it is difficult to find a basis to deny that they have inherent worth (once it is admitted that they have a good and on the assumption that nonsentient living natural entities have inherent worth).

Of course, all these puzzles are generated on the premise that at least some nonsentient, living, natural organisms have inherent worth. The puzzles evaporate if that premise is false.[7] Many people certainly believe, however, that nonsentient living things, such as trees and people in permanent vegetative states, have a good of their own.

So what are we to make of these puzzles? What is the significance of these puzzles to the ethics of synthetic biology? How important is it to resolve them and thus clarify the moral status of synthetic organisms?

The moral status of synthetic organisms is most relevant to how they are treated once they exist. If they have inherent worth, then they are not mere things, and one must ask whether, in the research or post-research process, they are being considered and treated commensurate with their moral status. That said, so long as synthetic organisms are comparatively simple organisms such as bacteria and yeast, their moral status is at most comparable to that of naturally occurring microorganisms. As discussed above, either such organisms lack inherent worth, or their inherent worth makes no practical claim on us. For this reason it is appropriate (given the state of the field) that the vast majority of discussion on the ethics of synthetic biology has focused on risks and benefits, regulation, oversight, and malign use rather than on the moral status of the organisms themselves (Presidential Commission for the Study of Bioethical Issues 2010).[8] However, as synthetic organisms become more complex, their moral status could take on greater importance. Moreover, it may be that some types of research are incompatible with their moral status (again, assuming the existence of increasingly complex organisms) (Basl 2010). This will be particularly true if (or when) synthetic biology begins to modify or create novel entities with cognitive capacities. In that event, moral status questions will take on a new level of significance.

Notes

1. Not all products of human activity are artifacts. A collection of garbage, a footprint, and excess carbon dioxide in the atmosphere are the result of human activity, but they are unintended by-products. Because they are not intentional (in most cases), they are not artifacts, as we use the term here.
2. While synthetic organisms are on the artifactual end of the natural–artifactual life-form continuum, they are not at the limit. Artificial life-forms and artificial intelligences created from entirely nonbiological material would be still more artifactual (Sandler 2012).

3. Other varieties of intrinsic value are based on interest-independent considerations. For example, a painting might be intrinsically valuable because of its beauty or cultural significance; and a natural landscape might be intrinsically valuable because of its grandeur or wildness.

4. The basis of selection can be described in multiple, equally adequate, ways. For example, the heart was selected for because it contributed to survival and reproduction; it was selected for because it contributed to circulation (Goode and Griffiths 1995).

5. Sandler (2011) discusses whether artifactualness is a morally relevant property of living things with respect to several varieties of intrinsic value, of which inherent worth is one.

6. As argued in the section on Puzzle One, the symmetry puzzle, relational properties (etiological properties) are relevant to whether an entity has a good and what the good consists in. It is for this reason that qualitatively identical entities with different goods are possible. However, the question here is not whether an entity has a good (or the role of extrinsic properties in determining that good), but whether, if it does have a good, extrinsic properties are relevant to whether its good should be considered.

7. For some arguments that these organisms do have inherent worth, see Goodpaster 1978; Taylor 1989; Varner 1998; and Sandler 2007. For objections, see Singer 1977; Regan 1983; and Feinberg 1963).

8. The other aspect of the ethics of synthetic biology that has drawn considerable attention concerns the ethics of crossing species boundaries or creating novel species.

References

Basl, John. 2010. State neutrality and the ethics of human enhancement technologies. *American Journal of Bioethics* 1 (2):41–48.

Boorse, Christopher. 1976. Wright on functions. *Philosophical Review* 85 (1):70–86.

Cahen, Harley. 2002. Against the moral considerability of ecosystems. In *Environmental Ethics: An Anthology*, ed. Andrew Light and H. Rolston III. Blackwell.

Cummins, Robert. 1975. Functional analysis. *Journal of Philosophy* 72:741–764.

Feinberg, Joel. 1963. The rights of animals and future generations. *Columbia Law Review* 63:673.

Feldman, F. 2004. *Pleasure and the Good Life: Concerning the Nature, Varieties and Plausibility of Hedonism.* New York: Oxford University Press.

Gibson, Daniel G., John I. Glass, Carole Lartigue, Vladimir N. Noskov, Ray-Yuan Chuang, Mikkel A. Algire, Gwynedd A. Benders, et al. 2010. Creation of a bacterial cell controlled by a chemically synthesized genome. *Science* 329 (5987): 52–56.

Goode, R., and P. E. Griffiths. 1995. The misuse of Sober's selection for/selection of distinction. *Biology and Philosophy* 10 (1):99–108.

Goodpaster, Kenneth. 1978. On being morally considerable. *Journal of Philosophy* 75:308–325.

Griffin, James. 1988. *Well-Being: Its Meaning, Measurement, and Moral Importance.* New York: Oxford University Press.

Hursthouse, Rosalind. 1999. *On Virtue Ethics.* Oxford: Oxford University Press.

Kawall, Jason. 2008. On behalf of biocentric individualism: A reply to Victoria Davion. *Environmental Ethics* 30:69–88.

Millikan, Ruth Garrett. 1989. In defense of proper functions. *Philosophy of Science* 56 (2):288–302.

Millikan, Ruth Garrett. 1999. Wings, spoons, pills, and quills: A pluralist theory of function. *Journal of Philosophy* 96 (4):191–206.

Mitchell, Sandra D. 1993. Dispositions or etiologies? A comment on Bigelow and Pargetter. *Journal of Philosophy* 90 (5):249–259.

Neander, Karen. 1991. Functions as selected effects: The conceptual analyst's defense. *Philosophy of Science* 58 (2):168–184.

Neander, Karen. 2008. The teleological notion of "function." *Australasian Journal of Philosophy* 69 (4): 454–468.

Nozick, Robert. 1974. *Anarchy, State, and Utopia.* New York: Basic Books.

O'Neill, John. 2003. The varieties of intrinsic value. In *Environmental Ethics: An Anthology*, ed. Andrew Light and Holmes Rolston III. Malden, MA: Blackwell.

Presidential Commission for the Study of Bioethical Issues. 2010. *New Directions: The Ethics of Synthetic Biology and Emerging Technologies.* Washington, D.C.: Presidential Commission for the Study of Bioethical Issues.

Regan, Tom. 1983. *The Case for Animal Rights.* Berkeley: University of California Press.

Ro, D., E. Paradise, M. Ouellet, K. Fisher, K. L. Newman, J. M. Ndungu, et al. 2006. Production of the antimalarial drug precursor artemisinic acid in engineered yeast. *Nature* 440:940–943.

Sandler, Ronald. 2007. *Character and Environment: A Virtue-Oriented Approach to Environmental Ethics.* New York: Columbia University Press.

Sandler, Ronald. 2011. Is artefactualness a value-relevant property of living things? *Synthese* 185 (1):89–102./

Sandler, Ronald. 2012. *The Ethics of Species.* Cambridge University Press.

Sandler, Ronald, and Luke Simons. 2012. The value of artefactual organisms. *Environmental Values* 19:43–61.

Singer, Peter. 1977. *Animal Liberation.* London: Paladin.

Singer, Peter. 1989. All animals are equal. In *Animal Rights and Human Obligations*, ed. Tom Regan and Peter Singer. Englewood Cliffs, NJ: Prentice Hall.

Streiffer, Robert, and John Basl. 2011. Applications of biotechnology to animals in agriculture. In *The Oxford Handbook of Animal Ethics*, ed. T. Beauchamp and R. Frey. Oxford University Press.

Taylor, Paul W. 1989. *Respect for Nature: Studies in moral, political, and legal Philosophy*. Princeton: Princeton University Press.

Varner, Gary. 1998. *In Nature's Interest*. Oxford: Oxford University Press.

Wright, Larry. 1973. Functions. *Philosophical Review* 82:139–168.

Wright, Larry. 1994. Functions. In *Conceptual Issues in Evolutionary Biology*, 2nd ed., ed. Elliott Sober. Cambridge, MA: MIT Press.

6

Synthetic Bacteria, Natural Processes, and Intrinsic Value

Christopher J. Preston

Today's synthetic biology, just like traditional biotechnology, raises important questions about the moral significance of its products; and questions about the "use" values and disvalues of synthetic organisms are especially prominent among these. It is, after all, the uses—and, by extension, the markets—for these new bacterial organisms that typically drive research into their production. Also significant, however, are questions about the intrinsic (or inherent) value of bacteria produced through synthetic means.[1] If the public's reaction to older forms of biotechnology is any indication, sentiments about these intrinsic values and disvalues are widespread, wielding a greatly underestimated, popular power.

Intrinsic-value arguments initially appear soft and somewhat vague, lacking the hard, pragmatic angle that characterizes arguments about instrumental benefits—such as energy production and medicinal uses—or dangers such as physiological harm. They do, however, have an emotional power that can be equally compelling. In the realm of traditional biotechnology, the intrinsic-value arguments about "playing God" and acting "unnaturally" were arguably as much responsible for the social movement against genetically modified crops in Europe as were worries about actual harm to ecosystems and traditional seeds. Although the synthetic biology pioneer Drew Endy has dismissed such negative reactions to synthetic biology as "superficial and embarrassingly simple," the evidence suggests that these kinds of intuitions have considerable reach and power.[2] For this reason, scoping the moral ground for the products of this new technology must involve in-depth discussion of the intrinsic value of synthetic organisms (Macilwain 2010).

There are, of course, numerous and varied research agendas in synthetic biology and numerous different potential products. This chapter focuses on the synthetic bacteria created in the J. Craig Venter Institute lab in May 2010. These bacterial cells contain a genome modeled on the genes found in *Mycoplasma mycoides*, synthesized in the laboratory entirely from constituent chemicals and inserted into a *M. capricolum* cell acting as host (Gibson et al. 2010). The Venter Institute's success in having the synthesized *M. mycoides* DNA take over the operation of the *capricolum* host was heralded the "world's first synthetic cell."[3]

Intrinsic and Instrumental Values

The distinction between instrumental and intrinsic value has a long history in the environmental ethics literature (Naess 1973; Rolston 1975; Regan 1981; Rolston 1988).[4] The distinction serves to differentiate between values found in nature that serve a human need or purpose, and values thought to inhere in nature independent of any human desires or interests. The reason the distinction has been so prominent is that, from the start, many held that *only* an ethic subscribing to the intrinsic value of nature was a truly *environmental* ethic. John O'Neill, for example, began one survey article with the claim that "to hold an environmental ethic is to hold that non-human beings and states of affairs in the natural world have intrinsic value" (O'Neill 1992, 119).[5] Only an ethic that refused to let value in nature depend entirely on human needs and interests was a genuinely environmental (and nonanthropocentric) ethic. In the minds of many environmental philosophers, intrinsic values became the litmus test for a true moral commitment to the earth itself.[6]

Although the reference to human interests initially appears to provide a fairly clear-cut distinction between two types of ethic, environmental philosophers found a great deal of complexity contained in the question of what exactly the term "intrinsic value" meant. O'Neill (1992), for example, distinguished three different senses in which the term gets used. The first captures the idea that the value of something is noninstrumental, the second that its value is entirely dependent on the nature of the object itself (and not on its relations to other things), and the third involves the metaethical claim that an object's value does not depend on the evaluative attitude of a subject (i.e., its value is not subjective). Keekok Lee made a distinction between value "enacted mutely" and value "recognized articulated" (Lee 1996, 299), the former found in organisms such as trees and bacteria, the latter in humans alone. While distinctions such as these add a whole archipelago of complexity to the discussion of intrinsic value, all the different senses of the term embrace the idea that humans should take nature's values into consideration *regardless of human interests*.

Answering the question of whether synthetic bacteria have intrinsic value or disvalue may provide guidance on how—or even whether—to proceed with their creation. Synthetic bacteria are clearly expected to have great instrumental value, but do they also have intrinsic value? Or is it possible they are intrinsically disvaluable? The answers to these questions could cut many different ways. Perhaps synthetic bacteria have enough intrinsic value to provide reasons to create them independent of the uses they might provide for humans.[7] Alternatively, an existing synthetic organism's intrinsic value could earn it protection from harm. On the opposite extreme, the intrinsic value of natural processes could be such that the creation of synthetic bacteria might pose a grave threat to existing natural value. This latter worry—appearing in the form of claims about acting unnaturally, crossing a moral line, or playing God—was common in the public debates about genetically modified organisms in the 1990s, and it can be expected to reappear in today's debates about synthetic organisms. At the very least, an investigation into the intrinsic value of synthetic organisms is necessary to determine whether a synthesized bacterium should always be viewed as merely a tool for some human purpose, or whether it has value as an organism in its own right.

Intrinsic-Value-Based Challenges to Synthetic Biology

Over the last few years, several philosophers have cautioned against synthetic organisms with arguments about the threat they pose to the intrinsic value of natural processes. In these accounts, the degree of manipulation of nature achieved by synthetic biology is said to be intrinsically wrong on the basis of the natural values being compromised. Joachim Boldt and Oliver Müller, for example, without explicitly utilizing the natural-value language, argue that the shift from the manipulation of genomes in traditional genetic modification to their creation in synthetic biology raises a host of new ethical issues (Boldt and Müller 2008; Schark 2012). They draw attention to the fact that living organisms have until now enjoyed a category of being entirely separate from machines. The blurring of the distinction between life and machine represents the crossing of an important threshold, one at which they urge scientists to show caution. The idea that living organisms can be artifacts, tailored entirely to human design, not only does something to how we value nonhuman nature, they suggest, but also risks having knock-on effects upon the way we value human life.

I also pressed an intrinsic-value angle in a paper raising the possibility that synthetic biology crosses “a line in Darwin's sand” (Preston 2008). Unlike traditional biotechnology, synthetic biology severs the link that living organisms have always maintained to their historical evolutionary past. In traditional biotechnology, the vast majority of the DNA in a modified organism is a product of descent with modification (or vertical transfer), with only a handful of genes altered through recombinant DNA technology.[8] In a fully synthetic organism, by contrast, all of the DNA is synthesized rather than inherited. Even though the minimum genome forming the blueprint of the synthetic organism may be modeled on a naturally occurring organism, the fact that the minimum genome is then synthesized entirely in the lab means that there is no physical, causal link between the manufactured organism and any ancestors. Since the concept of intrinsic value in environmental ethics often leans on some aspect of causal history between the present organism and the evolutionary past (Elliot 1982), severing the link presents “a clear basis for a deontological argument against synthetic biology” (Preston 2008, 34).

Recent papers by Gregory Kaebnick (2009) and Ron Sandler and Luke Simon (2012) have questioned the strength of these principle-based arguments. Two lines of questioning can be identified. The first asks whether it is plausible to suggest that synthetic organisms somehow interfere with the value of natural processes. The second asks whether synthetic organisms have their own intrinsic value, regardless of how they stand in relation to natural processes. (See also chapters 4, by Basl and Sandler, and 5, by Bedau and Larson, in this volume.) In the following discussion, the issue at stake is not the ecological or biological consequences of the introduction of synthetic organisms. Rather, it is the intrinsic values those organisms embody—or threaten—that matter.

The Impact of Synthetic Bacteria on Natural Processes

Both my own and Boldt and Müller's papers suggested that when a certain threshold is crossed, a number of widely appreciated values associated with natural processes come under threat. For my own part, articulating these threats was a way to try to make sense of reactions to synthetic organisms—including my own reactions—that involved unease or discomfort. The validity of these worries can best be evaluated by clarifying what exactly about natural processes the creation of synthetic bacteria does—or does not—threaten.[9]

The idea that synthetic biology crosses a "Darwinian line in the sand" suggests that synthetic biology somehow lays down a challenge to Darwinian theory. Despite the fact I described Darwinism in the 2008 paper as being "usurped" (p. 24), "departed from" (p. 34), and "superceded" (p. 35), the evolutionary processes Darwin described obviously do not abruptly cease to occur when synthetic organisms are created. The basic Darwinian process of descent with modification will continue despite the creation of synthetic organisms. In fact, since the first synthetic bacterium created in the Venter lab is already reproducing, the process is now operating on synthetic genomes as well. The idea that synthetic biology usurps Darwin is true only in the sense that lab-built organisms will henceforth exist that inherited none of their actual DNA through Darwinian processes. In those specific organisms, Darwinian explanations of the physical origin of an organism's DNA will play no part. It is in this very restricted sense that Darwinism is usurped. So the threat to Darwinian processes is, at best, simply a threat to its ubiquity as an explanation for the genomes of existing organisms.

A second qualification is that, while Boldt and Müller are correct to suggest that synthetic biology removes some of the existing clarity in the distinction between "life" and "machine" (Boldt and Müller 2008, 388), and while it might be true that synthetic biology creates "a more fundamental kind of biotic artifact" (Preston 2008, 35), synthetic organisms do not necessarily undercut the distinction, fundamental in environmental ethics, between nature and artifact. According to the traditional Aristotelian characterization, artifacts are things that do not exist by nature but are products of art or craft. When an object is produced by art or craft, according to Aristotle, the "principle [of its own production] is in something else external to the thing (Aristotle, *Physics*, 192b29–30). That "something else" is the artist or craftsman.

Despite the fact that a synthetic bacterium is a very unusual type, there does not appear to be any ambiguity about whether it is an artifact according to this Aristotelian criterion. Clearly it is. A synthetic organism, like all artifacts, comes into being as a result of human intention (see chapter 5 in this volume). Synthetic biologists may create organisms (or mimic existing ones), but they do not create "nature" in the pure Aristotelian sense.

The artifacts synthetic biology creates *are* different in important ways from anything to emerge in biotechnology thus far. Even though there have been organisms before that contain a portion of human intention and design (Pekingese dogs, for example, and Bt cotton), never before have there been organisms with genomes *entirely* the product of intentional human manufacture.[10] Synthetic organisms are thus an entirely new kind of organism *and* an entirely new kind of artifact. However, the nature/artifact distinction—and whatever conceptual usefulness it retains for the environmental community—does not appear to be under threat.[11]

A third important clarification (connected to the second) is that synthetic organisms do not undercut the agenda of environmentalists. In *The Natural and the Artifactual*, a book about what technology might do to the category of the natural, Keekok Lee worried that some technologies (such as nanotechnology and biotechnology) were "nature-replacing" in the sense that they literally replaced natural objects with artificial ones. She saw these technologies as paving the way for the elimination of natural kinds, threatening to "systematically transform naturally occurring beings (whether biotic or abiotic) to become artifactual ones" (Lee 1999, 1). In my paper, I judged that the real threat is posed not to the entire material fabric of nature (as Lee thought) but to the idea of biological organisms as archetypes of the natural. In the face of the challenge of artifactual organisms, "many positions in environmental ethics may need to be rethought" (Preston 2008, 37).

These worries, however, are in both cases exaggerated. Short of the now largely discredited fears of runaway ecophagy at the hands of gray or green goos, a few billion synthetic bacteria present no such risk.[12] To counter Lee's complaint, all one needs to do is point out that the intention with synthetic biology is not to replace all natural organisms with synthetic organisms. Synthetic biology simply adds synthetic organisms to the complement of the world's living beings for the purpose of serving selected goals. Environmentalism will still very much be about protecting polar bears, bull trout, and endangered songbirds. In fact, one might suggest—a little tongue-in-cheek—that synthetic organisms can provide a boost to global biodiversity by adding new biological types to the list of the world's organisms. It is not the intention of synthetic biologists to replace the species environmentalists seek to protect. The natural values that environmentalists treasure—and indeed the very concept of the natural world as something to cherish and protect—will remain largely intact.

In light of this examination of the alleged threats to Darwinism, to the nature/artifact distinction, and to the concept of nature, it is not clear what, if any, deontological arguments against synthetic bacteria follow from the conceptual challenge they present to natural processes. While

Boldt and Müller offer a caution against the crossing of the threshold between life and machine, they do not actually show why the threshold should not be crossed, apart from suggesting a certain feeling of discomfort and the possibility that synthetic biology is a slippery slope into the devaluation of life. One might point out to Boldt and Müller that the history of technological development is one of thresholds surpassed. Thresholds such as wild/domesticated, organic/conventional, and in utero/in vitro have all been crossed, with no moral argument ever proving decisive against the new technology. Certainly some of these thresholds have been historically interesting and some have also created considerable unease.[13] Perhaps the threshold between life and machine is another threshold worth noting, but one whose passage will not cause lingering cultural malaise.

The case is similar with my own claim about crossing a Darwinian line in the sand. Even if the line between organisms that have biological ancestors and those that have no ancestors in nature is indeed being crossed by the products of synthetic biology, this may be another line that has some historical significance, but no moral significance. References to "thresholds" and "lines in the sand" may be helpful for explaining the sense of discomfort and squeamishness created by synthetic organisms in some people, but may not amount to any reason to prohibit them.

Kaebnick (2009) examines the deontological arguments in the Boldt and Müller and the Preston papers and finds them all inadequate. He claims one would need a particular metaphysical account of reality (probably supported by a certain type of theology) to sustain the view that it is intrinsically wrong to manipulate nature through synthetic biology (Kaebnick 2009, 1107). Additionally, one would need arguments to show that any moral concerns raised by this type of manipulation of nature could not be outweighed by the benefits the products of synthetic biology promise. Cheap energy and effective medicines might end up being a good trade-off for crossing a line in the sand. Kaebnick further suggests that, from a policy perspective, the consequentialist arguments rather than the deontological ones are most important. Actual harms, such as inadvertent releases of synthetic organisms into the environment, biosecurity threats, and new pathogens synthesized in the lab, are the worries that should be taken seriously by policy makers, not the arguments suggesting that synthetic organisms somehow threaten natural values. The discomfort created in some quarters by synthetic biology (or the "yuck factor") is not, on its own, a decisive moral argument. There may be little wisdom in repugnance (Kass 1997). Even if synthetic bacteria create moral squeamishness in some, that squeamishness may be a legitimate target for the simple instruction, "Get over it!"[14]

The Intrinsic Value of Synthetic Bacteria

If the arguments drawing on moral squeamishness about their impact on natural processes are not decisive, then a second line of inquiry is to ask whether one can identify intrinsic value or disvalue in synthetic organisms themselves. If synthetic bacteria have their own intrinsic value and this value is added to the instrumental values they promise, then the moral calculus could swing decisively in favor of their creation and away from whatever discomfort they generate. On the other hand, if synthetic bacteria possess intrinsic disvalue, then the equation looks different. What, then, are the arguments for the intrinsic value of synthetic organisms?

One intriguing line of argument, mapped out by Ron Sandler and Luke Simon (2012) and by Basl and Sandler (in chapter 5), is that synthetic organisms possess intrinsic value on grounds similar to those on which natural organisms possess it. Synthetic organisms have intrinsic value because, like natural organisms, they are "teleologically organized"—that is, their "parts, processes, and operations are organized . . . for reasons pertaining to certain ends, such as survival, self-maintenance, and reproduction" (Sandler and Simon 2012, 50).[15] In chapter 5, Basl and Sandler hang the reality of this teleology on the fact that the different genetic parts of the synthetic organism are selected with certain functions in mind. Contra Rolston, Elliot, and other environmentalist positions described above, the question of the organism's intrinsic value does not depend on any connection to evolutionary history. While some thinkers will find value in this connection, it is never, according to Basl and Sandler, the main basis for an organism's intrinsic value. Bedau and Larson share Basl and Sandler's view, arguing similarly in chapter 4 of this volume that there are many other types of intrinsic values an organism possesses that have no connection to history or origins.

The core idea underpinning the teleological argument for intrinsic value is that if an organism has the ability to self-direct toward its own maintenance and survival, then it has a well-being that can be thwarted or harmed. "Even after its designers, creators and users fall away," state Sandler and Simon, "[the organism] still strives to promote its ends and accomplishing those ends is beneficial to it" (p. 52). Such teleologically organized beings have "goods of their own" that are independent of other beings.

The argument makes the major part of an organism's value *capacities based*. Whatever value it has, that is, must depend on some internal capacity of the organism rather than depending on the organism's relations to anything (or anyone) else. So the fact that it was constructed in a lab rather than in the natural environment, or that its genome was designed by humans rather than being the product of evolutionary inheritance, can have no bearing on the organism's capacities-based intrinsic worth. If its value were based on anything other than the organism's own capacities, then the value would be in some way *extrinsic* to the organism, which is precisely what intrinsic value cannot be.[16] On this account, a teleologically organized artifactual organism produced by synthetic biology possesses intrinsic value in the same way that a natural one does. As Bedau and Larson put it in chapter 4, they both have "biological needs." They both function in a way that protects their own well-being.

The idea that moral considerability should be capacities-based has a long and venerable history in ethics. In utilitarianism, for example, the capacity to suffer pain and enjoy pleasure is the criterion for moral standing. In Kantianism, a being has dignity (and hence, intrinsic worth[17]) if it is rational and has the capacity to act on its own conception of the moral law. Basing moral considerability on the individual's own capacities rather than on anything extrinsic to the individual is in fact central to the liberal tradition. It protects against inequality by embracing the idea that beings with similar capacities are due similar moral regard. The question for current purposes is whether this fundamental principle of liberalism is appropriately extended to synthetic organisms.

Supporters of synthetic biology employing this teleological account join numerous environmental ethicists in trading the restrictive historical accounts of moral significance for more permissive positions. As an instructive example, Paul Taylor (1986) took what was essentially the Kantian intuition about intrinsic value and made it more permissive by suggesting that any "teleological center of life" deserved moral consideration. To be a teleological center of life, Taylor stipulated, an organism had to be "a unified, coherently ordered system of goal-oriented activities that has a constant tendency to protect and maintain the organism's existence" (Taylor 1986, 122). "To say that [an organism] is a teleological center of life," Taylor added, "is to say that its internal functioning as well as its external activities are all goal-oriented" (p. 121). For Taylor, plants and animals (in addition to humans) have moral worth on these grounds. If Taylor is right, then it certainly looks like a successful synthetic organism would meet the same criteria.

The case of synthetic organisms, however, is arguably not so straightforward. No doubt, a successful synthetic organism is in certain respects teleologically organized and strives to maintain its own existence. But a synthetic organism is arguably not teleologically organized in quite the same way as a naturally occurring one. It has a slightly different kind of teleological character, making the leap from the intrinsic value of naturally occurring organisms to the equal intrinsic value artifactual organisms problematic. To illustrate this difference, it proves helpful to borrow some philosophical language from an unexpected quarter.

The Four Causes in Nature and in Artifacts

While it might seem odd to employ Aristotelian language to pursue an intuition about one of the most recent developments in contemporary biology, the thorough nature of the causal explanations demanded by Aristotle turns out to be helpful in peeling apart the ways a synthetic organism might be morally different from a naturally occurring one. According to one of the most memorable discussions in Aristotle's *Metaphysics*, in order to fully understand anything (natural or artifactual), one needs to give an account of its four causes—material, moving, final, and formal. In the case of artifacts, three of those four causes are imposed from without. So, for example, a carved wooden bowl's *moving* cause is the craftsman who fashions it on a lathe, its *formal* cause is the bowl design that the craftsman models her work on, and its *final* cause is the purpose for the sake of which the bowl is being constructed (e.g., to hold food or water). All of these three causes reference considerations that originate externally.

The remaining cause—the bowl's material cause—is the wood of which the bowl happens to be composed. Insofar as the bowl is made from wood, some of its behavior appears to be generated internally (for example, the bowl will eventually rot, it can burn, and so on). However, the bowl might have been made of tin or clay, so the pace of its eventual decomposition and its flammability is not part of the nature of a bowl per se, but reflects what Aristotle calls a "concomitant attribute" (*Physics* II, 192b31), given that the craftsman happened to choose wood for the bowl's material. It is clear, then, from a look at these four causes that a great deal of what we need to know to fully understand the bowl hinges on factors external to the bowl itself.

In nature, by contrast, none of the causes are sourced from without. For natural objects, the four causes are all part of the internal nature of the thing. In Aristotle's analysis, the final and the formal cause of naturally occurring objects usually overlap and point toward the nature of the organism itself. In a young oak tree, the sapling does what it does at this point of the season because it is internally driven to assume the typical form of 'oak tree' (its formal cause). Moreover, this is its behavior because being an example of 'oak tree' is its natural purpose (its final cause).[18] Aristotle then often connects the moving cause of a natural organism to its form, since the source of the motion contained in any organism is typically another organism of the same kind (the actual oak tree producing the acorn that became the sapling in question).[19] The material cause of the sapling is the wood, made up of the elements of which it is composed. The four causal explanations are all different from the case of the bowl. To explain changes undergone by the oak sapling, one needs only to know things about oak trees and about how (and where) they grow. One need not know anything about the intentions of any other agent.

These differences in the causal explanations of natural substances and artifacts prove to be telling differences for synthetic bacteria. They help explain how a synthetic organism's teleology is different from that of a naturally occurring one. Though both are teleologically organized, they are teleologically organized in different ways.

The complex fourfold nature of Aristotle's explanatory framework makes it necessary to give different causal analyses of synthetic bacteria at different stages of their existence. The very first synthetic genome produced in the lab requires one causal analysis just before it is "booted up" in its host cell. Once the DNA starts functioning in its host cell, a second causal analysis is necessary. When this synthetic bacterium starts reproducing autonomously, its descendants require a third causal analysis. While the first two of these causal analyses are interesting in their own right,[20] the most pertinent analysis for our purposes is the third one, the analysis of the descendants. These bacterial organisms will be by far the most numerous over time, and they will be the ones performing the tasks for which these synthetic organisms will are designed.

Once a synthetic organism is reproducing in the laboratory—something that *M. mycoides* JCVI-syn1.0 is already doing in Rockville, Maryland—the causal analysis takes an unusual turn that Aristotle could not possibly have anticipated. The material cause of this second (and subsequent) generation synthetic bacterium is relatively straightforward and similar to the material cause of any naturally occurring organism. The descendants of *M. mycoides* JCVI-syn1.0 are reproduced from the existing bacterium through the process of binary fission. The material causes of their behavior are the elements taken up from the ambient environment and assembled in a certain order by the bacterial DNA. For the moving cause, the analysis also remains similar to that of any naturally occurring organism. As with the oak tree, the moving cause is the parent organism, in this case *M. mycoides* JCVI-syn1.0.

As for the formal and final causes, the analysis is far from straightforward. In order to accord with the analysis of the oak tree and other teleologically organized organisms, we must follow Aristotle and say that the formal and final causes of this synthetic bacterium overlap. A descendant bacterium's purpose (final cause) is to become a *M. mycoides* JCVI-syn1.0, and the form that guides its growth (formal cause) on the way to that end is also *M. mycoides* JCVI-syn1.0. So far the explanation has paralleled that of a naturally occurring organism, reflecting the fact that a synthetic bacterium appears to have a good of its own (Sandler and Basl) and biological needs (Bedau and Larson). Unfortunately, however, the job of specifying the formal and final causes is clearly not complete. It would be an error not to add that the final cause of these bacteria, the purpose for which they exist, is to perform a certain function chosen by their designers. It would also be an error to fail to point out that their formal cause remains the particular blueprint chosen by the researchers as the basis for *M. mycoides* JCVI-syn1.0 and stored on the laboratory computer. The necessity of these additional explanations means that the formal and final causes of synthetic organisms turn out to be bifurcated. This bifurcation is a consequence of the fact that a synthetic organism is, in some important sense, simultaneously both an artifact *and* an organism.

The bifurcation of formal and final causes for synthetic organisms is irreducible. To say only (as one would with a naturally occurring organism) that these bacteria are teleologically organized to self-maintain and reproduce would not provide a full understanding of how they came into being. Similarly, to say only (as one would with an artifact) that these bacteria are organized to perform a certain function according to a design on a computer would be to leave out half the story. Neither explanation would provide, as Aristotle requests, "the 'why' of it . . . as regards both coming to be and passing away and every kind of physical change" (*Physics* II, 194b20–22). To understand the "why" of these bacteria, one still needs to know something about bacteria and DNA, about synthetic biology as a research area, about the complexity of synthesizing a genome, and about a whole suite of design decisions made by the technicians involved. In other words, while some of the explanation of the organism's behavior is now certainly internal (belonging to the functional organization of its parts), other parts of the explanation remain irreducibly external. At some level, the technicians in the lab will always remain responsible for the existence of these very complex, self-replicating, biological machines.[21]

Assessing the Intrinsic Worth of Synthetic Organisms

With this Aristotelian account of the difference between a naturally occurring organism and a synthetic organism in hand, one can assess anew the intrinsic value of synthetic organisms. The suggestion above had been that the intrinsic value of an organism was a consequence of its teleological organization. Teleological organization corresponds here to the discussion of final causes and the functions different parts of the organism serve. What we have seen is that in the case of a naturally occurring bacterium the final cause is singular, whereas in the case of a synthetic one it is irreducibly bifurcated. Since the teleological analysis of a synthetic bacterium is clearly different from the teleological analysis of a naturally occurring organism, it seems unlikely that their intrinsic value can be the same. The remainder of this chapter posits, in three arguments, that the intrinsic value of a synthetic organism is not only likely to be *different* but also likely to be *diminished* relative to the intrinsic value of a naturally occurring one.

Argument One: Teleological Independence

As a preliminary point, we should note that it is widely held that not everything teleologically organized has intrinsic value.[22] Thermostats are teleologically organized. They are created for the purpose of turning things on and off when a certain temperature is reached. But while thermostats clearly have instrumental value, it seems unlikely that they have intrinsic worth.

Obviously thermostats are different from synthetic bacteria; thermostats are not organisms. However, it is not always the case that simply being an organism is reason enough for something to have intrinsic value. In his detailed analysis of the complicated language of intrinsic values, John O'Neill held that simply having teleological organization does not automatically mean that an organism possesses intrinsic value; a point also acknowledged by Sandler and Basl in chapter 4. One could think that there is no intrinsic value to the AIDS virus, even though the virus itself is teleologically organized and has a good of its own. O'Neill formally captured this point when he stated that "we can know what is 'good for X' and relatedly what constitutes 'flourishing for X' and yet believe that X is the sort of thing that ought not to exist and hence that the flourishing of

X is just the sort of thing we ought to inhibit" (O'Neill 1992, 131). So it is at least theoretically possible that some teleologically organized things have less intrinsic value than others, and perhaps zero value.

Another pertinent preliminary point made by O'Neill is his denial that an organism's intrinsic value must hinge only on its nonrelational properties (O'Neill 1992, 124). To assume it does is to confuse the sense of intrinsic that means noninstrumental with the sense that means nonrelational. Organisms can have noninstrumental intrinsic value that *is* relational. O'Neill points out, for example, that organisms can gain (or lose) intrinsic value in the noninstrumental sense on the basis of relational properties such as rarity.

This matters in the case of synthetic bacteria because a key part of their teleological organization—the criterion on which we are supposed to be basing their intrinsic value—*is* relational, namely the intentions of those who designed them. A synthetic bacterium's bifurcated teleological organization means that part of its final cause is relational (concerning the organism's designated purpose) and part of it is nonrelational (concerning its own self-maintenance).

Although it cannot here be shown conclusively, there is reason to suspect that these relational aspects of a synthetic bacterium's teleological organization might function as a value diminisher. Part of the reason we do not think a teleologically organized thermostat possesses intrinsic value is that its good—the ability to switch on and off at the right temperature—is not its own, but a good chosen by the technicians who designed it. Since the formal and final causes of the synthetic bacterium remain, irreducibly and permanently, connected to the technicians who designed them, the good of the synthetic bacterium is never entirely its own.[23] Since one of the features environmentalists most admire about natural organisms is their independence (both in origin and in function), the lack of independence in a synthetic bacterium's final cause may serve to diminish its value relative to a natural organism whose final cause is not bifurcated.

Argument Two: Prior Cause

There is more to be said about the implications of synthetic bacteria having two final causes. The problem is not only a lack of teleological independence, but also the causal priority found within its bifurcated teleology. For current purposes, the two final causes of a synthetic bacterium might be characterized as an "organismal final cause" (attributable to the organism's autonomous functioning) and an "artifactual final cause" (attributable to the engineer's goals and design specifications). The former is wholly internal to the organism, the latter external. Arguably the artifactual final cause is, both temporally and conceptually, prior to the organismal final cause. Without the intentions of the lab technicians, *M. mycoides* JCVI-syn1.0 would never have existed at all. It came into existence precisely in order to demonstrate a certain scientific possibility, just as future synthetic organisms will come into existence to demonstrate the possibility of a minimal viable genome, or to demonstrate and perform the function of a carbon capturing bacterium, or a medically useful bacterium, and so on. None of these organisms would exist without their artifactual final cause.

If the artifactual final cause is always temporally and conceptually prior to the organismal final cause, then it is worth asking whether this might diminish the synthetic organism's intrinsic value. It seems that it might. Nonorganismal artifacts created for an instrumentalist purpose are most often presumed not to have intrinsic worth. This is why a broken thermostat is generally not something we feel a moral obligation to fix, at least not for the sake of the thermostat itself. The artifactual final cause is not typically thought to confer any intrinsic value on the thermostat. If the artifactual final cause is the prior cause of the two final causes for the synthetic organism, then it looks like this may serve to diminish its overall value.

To press this point further, consider the case of domesticated animals, which are also both artifacts and organisms. Even though they are not artifacts in quite the same way as synthetic bacteria are—they have not crossed the Darwinian line in the sand, and they retain ancestors from whom they inherited their genes—philosophers from John Muir (1874) to J. Baird Callicott (1980) have treated them as possessing diminished intrinsic value because they have been bred for traits deemed desirable to humans. For many environmentalists outside of the animal rights movement, the value of a Rocky Mountain bighorn sheep eclipses that of a sheep on the neighborhood farm.

Not everyone thinks this way about domesticated animals, of course. Animal rights advocates often consider domesticated animals to have identical moral significance to wild animals on the grounds of the equality of their capacities. However, the prevalence of poor treatment of domesticated animals for human consumption suggests that this view is not the dominant one. The fact that a particular steer has been bred for meat apparently diminishes the moral value many people are prepared to assign it. Because beef cattle, dairy cows, and egg-laying chickens serve a human purpose, they tend to be thought of as possessing less intrinsic value than wild bison, bighorn sheep, and wild turkeys.

Since the artifactuality of a synthetic bacterium (as measured by percentage of genes synthesized) far exceeds that of a sheep or a cow, the presumption that artifactual organisms contain less intrinsic value than nonartifactual organisms seems even more plausible in the case of synthetic bacteria than in the case of sheep. Short of an account of how the instrumentalist final cause of a synthetic bacterium might somehow "wash out" over time, a possibility the Aristotelian analysis above appears to doubt, the synthetic bacterium's intrinsic value may never match that of a naturally occurring organism.

Argument Three: Concomitant Attributes

As we have already seen, when Aristotle differentiates between nature and artifacts by distinguishing between causes that are internal and causes that are external, he also discusses causes that are contingent in particular artifacts. In the example of the wooden bowl, the fact that all wooden bowls will eventually rot is not part of the nature of bowls themselves but is due to the fact that these particular bowls happened to be made out of wood (something Aristotle called a "concomitant attribute"). Concomitant attributes of a thing are attributes that "are not always found together" with the thing (*Physics*, 192b27). There is nothing essential about bowls that requires them to be wooden. Many, indeed, are made of clay.

A similar point can be made about many of the synthetic organisms now envisioned. An organism might be created to serve as a carbon-capturing device, for example, but there is nothing about carbon-capturing artifacts that requires them to be organisms. Similarly, there is nothing about a particular medicine-producing system that requires it to be a synthetic bacterium. In fact, if the carbon capture or the medicine production could be done more efficiently using other means, then surely it would be. The quest to produce mechanical devices (known as "synthetic trees") that scrub carbon directly out of the atmosphere is currently further along than the quest to produce bacteria that perform the same function.[24] If it turns out these mechanical (nonorganismal) carbon scrubbers can do the job more efficiently than synthetic bacteria, then there will likely no longer be any need to develop a carbon-eating synthetic bacterium. Like the fact that a particular bowl happens to be wooden, the organismal nature of a device developed to capture carbon is a concomitant attribute of the technology, not a necessary one. It is incidental to the purpose for which the device is designed.[25]

In a section of *Physics* II that follows shortly after the discussion of concomitant attributes, Aristotle makes it clear that something happening by chance is posterior to something happening by nature. "Since nothing which is incidental is prior to what is per se," he says, "it is clear that no incidental cause can be prior to a cause per se" (*Physics* II, 198a6–9). The incidental fact that a carbon-capturing artifact might be biological cannot be prior to the fact that it effectively performs its instrumentalist function. It is more important that the artifact captures carbon than that it happens to be biological.

This discussion of concomitant attributes provides another reason to suspect that the sense in which synthetic bacteria are tools serving a particular purpose takes precedence over the sense in which they are autonomous organisms. Incidental attributes cannot be prior to essential attributes. It does not follow, however, that in all situations where organismal nature is a concomitant attribute, one is always entitled to treat something entirely as an instrument rather than an organism. The instrument that checks my subway ticket may be a person or a machine. Their personhood (or machine-hood) is a concomitant attribute of the ticket-checking function. If the ticket checker is a person, I must treat him as such rather than simply as a ticket-checking thing. However, the case of the person who is also a ticket-checker appears importantly different from the case of the synthetic bacterium that is also a carbon-capturing device. Ticket-checking is itself incidental to the ticket checker's existence as a person, whereas carbon capturing is not incidental to the bacterium's existence. Carbon capturing is an irreducible part of the final cause of the bacterium's existence.

Conclusion

These three arguments provide reason to doubt that synthetic bacteria have equal intrinsic value to similarly organized naturally occurring bacteria simply on the grounds that they both possess teleological organization. While it may previously have been the case that teleological organization was a good guide to the presence of intrinsic values in organisms, this may no longer be so. With the creation of synthetic organisms, teleological organization has become more complex. The Aristotelian analysis shows how synthetic organisms are teleologically organized in a way that is different from nonsynthetic organisms. There is reason to think that synthetic bacteria's having two irreducible final causes may make them morally different from naturally occurring bacteria. The irreducible and permanent artifactual final cause of synthetic bacteria may serve as a value diminisher relative to naturally occurring organisms. These conclusions largely follow from the fact that a synthetic bacterium remains irreducibly an artifact, or a tool manufactured to perform a particular purpose, in addition to (contingently) being an organism.[26]

That this whole discussion has been about bacteria may, in the final analysis, make the question of intrinsic value moot. As both Bedau and Larson and Sandler and Basl acknowledge in their chapters, we don't generally spend much time thinking about the intrinsic value of bacteria and other simple structures such as viruses. While readily acknowledging the instrumental value of these structures for their essential role in maintaining functional ecosystems and healthy physiologies, we usually ascribe intrinsic value only to more complex organisms. If bacteria in general have little or no intrinsic value, then we need not worry about the relative merits of synthetic versus natural organisms. This would certainly simplify the moral territory for the time being. If bacteria do have intrinsic value, however, there are reasons to believe that, when considering them on their own merit rather than based on the services they provide, we owe synthetic bacteria less. These considerations will become more important if synthetic biologists eventually create more complex organisms.

Notes

1. Talking about intrinsic value serves here as shorthand for a suite of deontological positions about the significance of nature, including some that do not expressly employ value language. Under this broad interpretation, to act "unnaturally" is to fail to respect the intrinsic value of nature. To "play God" is to ignore the special value of what God has created, and so on. What all these arguments have in common is a principle-based (i.e., nonconsequentialist) reaction to synthetic biology.
2. See Endy 2008. To illustrate the prevalence of this concern as it appears in agricultural biotechnology, Greg Kaebnick (2007) cites findings of a Pew poll suggesting that two-thirds of the public were "uncomfortable" with animal cloning even though less than half of them thought the products would present a threat to safety (Kaebnick 2007, 572). Preliminary social science inquiry into climate engineering strategies suggests that perception of the "naturalness" of a technology is an important factor in determining public support (NERC 2010). It seems quite probable that these initial responses (sometimes referred to as the "yuck factor") may often drive social movements against different biotechnologies even if it is the consequentialist arguments that end up most influencing public policy (Kaebnick 2009).
3. See Venter Institute press release, http://www.jcvi.org/cms/press/press-releases/full-text/article/first-self-replicating-synthetic-bacterial-cell-constructed-by-j-craig-venter-institute-researcher/.

4. This distinction mostly coincides with another important philosophical distinction between ethical positions that are deontological (Kantian) and those that are consequentialist (utilitarian). The intrinsic-value arguments of concern in this paper all occupy the deontological wing of ethics.

5. The use of the terms "intrinsic," "inherent," and "natural" value has not been consistent among environmental philosophers. At the request of the volume editor, "intrinsic" is used throughout.

6. This position has been challenged aggressively by "environmental pragmatists," "weak anthropocentrists," and others. That in-house debate will not concern us here.

7. The idea that it is morally good to bring a not-yet-existing organism into existence is one that has generated plenty of discussion in animal ethics (Singer 2001).

8. The role of horizontal (or lateral) transfer of genetic material through viruses and the uptake of mitochondrial DNA must also be acknowledged, for its role in evolution and for the challenge it presents to the familiar idea of descent with modification (Doolittle 2000). Most accounts of intrinsic value in the environmental ethics literature, however, rely on the assumption that the vast majority of DNA is inherited through vertical transfer, an assumption that may yet prove to be correct (Theobald 2010).

9. Some of these clarifications concern claims that were not clear enough in my original paper. Others involve modifications to my position that seem warranted in the light of (i) discussion with Ron Sandler, (ii) a meeting on Synthetic Biology at The Hastings Center in March of 2010, and (iii) consideration of the two papers discussed (Kaebnick 2009; Sandler and Simon 2012).

10. The synthetic organism created in the Venter lab in May 2010 is more or less a copy of an existing genome rather than a genome designed by humans. However, all of the genome's physical constituents were intentionally manufactured by technicians.

11. While the nature/artifact distinction is not under threat, the organism/artifact distinction clearly is. Some might not be concerned about this, suspecting that the threshold was crossed millennia ago with the domestication of wild animals. It seems, however, that a synthetic bacterium is artifactual to an unprecedented degree.

12. The gray goo worry was first popularized by Eric Drexler in his 1986 book *Engines of Creation: The Coming Era of Nanotechnology* (New York: Anchor).

13. This is particularly true for the line between in utero and in vitro fertilization.

14. The creation of discomfort among a significant section of the population—even if the discomfort rests on a shaky basis—can certainly form part of an argument against doing something. But for this partial argument to be decisive, one must also give reasons why the moral discomfort should not be outweighed by the moral advantages the technology is designed to provide. Numerous policy advances, from the abolition of slavery to the use of stem cells, have caused moral discomfort in some portions of the population.

15. I am amending Sandler and Simon's terminology to remain consistent with the rest of this chapter. What I am calling "intrinsic value," Sandler and Simon called "inherent worth."

16. O'Neill disputes this claim by pointing out that some relational (or extrinsic) properties of an organism, for example, its rarity, can contribute to its intrinsic value (O'Neill 1992, 124).

17. The language Kant used here is usually translated not as "intrinsic" but "inherent" worth.

18. For a thorough explanation of this point see Cooper 1982, 200–202.

19. *Physics* II, 198a24–25. It should be noted that occasionally Aristotle switches and identifies the moving cause with the material cause. Some explanations—an example Aristotle uses here is eye color—are not connected to any particular purpose but to the necessity of the matter out of which a particular iris is made (*Generation of Animals*, 778a32–b1).

20. The first and second analysis would go something like the following. In the instant prior to their insertion into the host cell, the object that is the not-yet-biologically-functioning-synthetic-bacterium-built-as-a-proof-of-concept has a causal analysis that is in many respects similar to that of the wooden bowl. It has, for example, as its moving cause the technicians in the Venter lab. Its material cause is the chemical soup of adenine, cytosine, guanine, and thymine out of which the DNA was synthesized. It formal cause is the map of the *Mycoplasma mycoides* JCVI-syn1.0 genome the technicians followed. Its final cause is to be a proof of concept to show that such an organism can be constructed. The analysis is conceptually similar to that for the bowl because, in the moment before the cell starts functioning as an organism, it is simply another artifact. As soon as the DNA inserted into the *Mycoplasma capricolum* host cell is "booted up" and takes over the function of the host cell, the situation changes. The cell is now doing things on its own and in some sense has become autonomous. The causal analysis accordingly increases in complexity. The moving cause seems still to be the technicians who put the cell together. The material cause is still the base chemicals originally inserted into the cell, but these chemicals are now gradually being replaced by chemicals taken up from the synthetic cell's ambient environment as it maintains and repairs itself. The formal and final causes are impossible to define with precision. The formal cause seems to be bifurcated. On the one hand the formal cause remains the blueprint of *M. mycoides* JCVI-syn1.0 stored in the Venter Institute's computers. On the other hand, since it is a self-maintaining organism, the formal cause has become the genetic makeup of this new type of organism, *M. mycoides* JCVI-syn1.0. We find a similar bifurcation for the final cause. On the one hand, the organism exists simply for the purpose of being a proof of concept. On the other, just as for the oak sapling, its purpose is to be an example of *M. mycoides* JCVI-syn1.0.

21. To illustrate the persisting significance of this external cause, imagine what would happen were something to go wrong. If these organisms escaped from the lab and wreaked ecological havoc, we would not view this as an unfortunate natural disaster. We would expect some accountability from those who built them.

22. This is identified in the course of "Puzzle Two" in chapter 5 by Basl and Sandler.

23. The point is not that some of the value synthetic bacteria provide is instrumental value. The point is that some of their final cause is irreducibly external.

24. "Artificial Tree Becomes Carbon Castle," *EnvironmentResearchWeb*, May 4, 2010, http://environmentalresearchweb.org/cws/article/news/42521.

25. An interesting exception here is a synthetic bacterium built as a proof of concept. If the artifact is built precisely to prove that a biological organism can be synthesized in the lab, then its biological nature is not accidental but necessary. Once the concept has been proven, future synthetic organisms will be built for more obviously instrumentalist purposes.

26. Similar considerations may apply to questions of the moral status of robots in the artificial intelligence debates (Russell and Norvig 2002).

References

Aristotle. *Physics*. Trans. R. P. Hardie and R. K. Gaye. New York: Random House, 1941.

Boldt, Joachim, and Oliver Müller. 2008. Newtons of the leaves of grass. *Nature Biotechnology* 26 (4):386–389.

Callicott, J. Baird. 1980. Animal liberation: A triangular affair. *Environmental Ethics* 2 (4):311–338.

Cooper, John M. 1982. Aristotle on natural theology. In *Language and Logos*, ed. Malcolm Schofield and Martha Nussbaum, 197–222. Cambridge: Cambridge University Press.

Doolittle, W. Ford. 2000. Uprooting the tree of Life. *Scientific American*, February, 90–95.

Elliot, Robert. 1982. Faking nature: The ethics of environmental restoration. *Inquiry* 25:129–133.

Endy, Drew. 2008. Life, what a concept! (Part III). *Edge* 237, February 19. http://www.edge.org/documents/archive/edge237.html.

Gibson, Daniel, John Glass, Carole Lartigue, Vladimir Noskov, Ray-Yuan Chuang, Mikkel Algire, Gwynedd Benders, et al. 2010. Creation of a bacterial cell controlled by a chemically synthesized genome. *Science* 329 (5987):52–56.

Kaebnick, G. 2007. Putting concerns about nature in context: The case of agricultural biotechnology. *Perspectives in Biology and Medicine* 50 (4):572–584.

Kaebnick, G. 2009. Should moral objections to synthetic biology affect public policy? *Nature Biotechnology* 27 (12):1106–1108.

Kass, L. 1997. The wisdom of repugnance. *New Republic* 216 (22):17–26.

Lee, Keekok. 1996. The source and locus of intrinsic value: A reexamination. *Environmental Ethics* 18 (3):297–309.

Lee, Keekok. 1999. *The Natural and the Artifactual: The Implications of Deep Science and Deep Technology for Environmental Philosophy*. New York: Lexington Books.

Macilwain, Colin. 2010. Talking the talk: Without effective public engagement, there will be no synthetic biology in Europe. *Nature* 465:867.

Muir, John. 1874. The wild sheep of California. *Overland Monthly* 12 (4):358–363.

Naess, Arne. 1973. The shallow and the deep, long-range ecology movement: A summary. *Inquiry* 16 (1):95–100.

National Environment Research Council. 2010. Experiment Earth? Report on public dialogue on geoengineering. http://www.nerc.ac.uk/about/consult/geoengineering-dialogue-final-report.pdf.

O'Neill, John. 1992. The varieties of intrinsic value. *Monist* 75 (2):119–138.

Preston, Christopher J. 2008. Synthetic biology: Drawing a line in Darwin's sand. *Environmental Values* 17 (1):23–39.

Regan, Tom. 1981. The nature and possibility of an environmental ethic. *Environmental Ethics* 3 (1):19–34.

Rolston, Holmes III. 1975. Is there an ecological ethic? *Ethics* 85:93–109.

Rolston, Holmes III. 1988. *Environmental Ethics: Duties to and Values in the Natural World*. Philadelphia: Temple University Press.

Russell, Stuart J., and Peter Norvig. 2002. *Artificial Intelligence: A Modern Approach* (2nd ed.) Englewood Cliffs, NJ: Prentice Hall.

Sandler, Ron, and Luke Simon. 2012. The value of artefactual organisms. *Environmental Values* 20 (4):43–61.

Schark, Marianne. 2012. Synthetic biology and the distinction between organisms and machines. *Environmental Values* 21 (1):19–41.

Singer, Peter. 2001. *Animal Liberation*. New York: Ecco.

Taylor, Paul. 1986. *Respect for Nature: A Theory of Environmental Ethics*. Princeton: Princeton University Press.

Theobald, Douglas. 2010. A formal test of the theory of universal common ancestry. *Nature* 465 (May 13): 219–222.

III

Values and Public Policy

7

Synthetic Biology and Public Reason

Jon Mandle

The developments associated with synthetic biology and other new genetic technologies raise profound questions in many different areas. Among others, they touch on issues of religion, metaphysics, and morality. Without in any way denigrating the importance of these deep and weighty issues, I am going to argue that the answers we give to these questions are largely independent of the answers we give to another set of concerns—those of public policy. Our answers to theological, to metaphysical, and in an important sense to moral questions about the technologies associated with genetic manipulation should not have any *direct* policy implications. I need to be especially careful with this last claim—about the independence of morality—because my claim is not that policy should not be influenced by moral concerns. On the contrary, I will argue precisely that it should be, but only by a certain limited class of moral considerations. And some of the moral and ethical issues raised by synthetic biology go beyond these limits.

I take my orientation here from the work of the philosopher John Rawls. Rawls never wrote about the ethics of genetic engineering. The closest he came was a few brief sentences that mention eugenics in passing (Rawls 1999).[1] Yet, I believe that we can apply and extend his approach to social justice in ways that will clarify some of the issues that these technologies raise. In developing what he calls a "political conception of justice," Rawls also articulates an account of *public reason*. This is the framework on which a democratic society ought to rely when deliberating over the legitimate use of its political power. I will start by explaining Rawls's account, before moving on to consider some implications for a society's deliberations concerning synthetic biology. Throughout, my emphasis will be procedural, so to speak. I will not be arguing that some particular regulation or policy is the correct one. Rather I will be examining what kinds of considerations and arguments are appropriate when making such political decisions.

The Political Conception of Justice

Rawls's idea of a political conception of justice starts by recognizing the deep and persistent diversity of basic religious, philosophical, and moral outlooks characteristic of modern societies. There are two key points to understanding how Rawls thinks about this diversity of comprehensive doctrines. First, many of these doctrines really do conflict with one another, and we should assume that there will not be any resolution of these especially deep disagreements. From the point of view of any one of these doctrines, other conflicting doctrines will appear to be wrong, false, incorrect, or misguided. Because they conflict, they cannot all be true or correct. At least some—maybe all—are false. It is a fundamental mistake to assume that these disagreements either are merely apparent, mask some deeper agreement, or will soon be transcended. Second, more dramatically, Rawls holds that at least some of these persistent disagreements are the result of the normal exercise of human reasoning. That is, at least some are not due merely to narrow, distorting and deforming ideologies, or failures of critical reflection, or objectionable selfishness. Some doctrines, and the resulting disputes, are undoubtedly the result of these and other factors. Such doctrines are unreasonable. But even putting aside these unreasonable doctrines, there will still be a diversity of conflicting but *reasonable* doctrines. Perhaps, if we all had exactly the same experiences, were raised in exactly the same way, assessed evidence in the same manner, and were all ideal reasoners with infinite time and patience to consider all the implications and consequences of our views, we would all reach the same conclusions and hold the same comprehensive doctrines. But that is not the human condition. On the contrary, the diversity of reasonable comprehensive doctrines is "the natural outcome of the activities of human reason under enduring free institutions" (Rawls 2005, xxiv).[2]

Rawls understands legitimate political power to be "the power of free and equal citizens as a collective body" (2005, 136). This more or less directly requires a democratic political structure. Anything else would obviously involve treating citizens as unequal in an important respect. But in order for the political actions of a society to be truly collective, it must be possible for the society's *reasons* for its actions and decisions to be shared as well. This does not mean that there must be unanimous agreement in order for a society's decision to be legitimate. But it does mean that all reasonable people must be able to recognize that the considerations cited in favor of a decision do, in fact, support it. Other considerations may certainly tell against the decision. What remains possible, and is often likely, is that there may be disagreement concerning the strength and relative weights of these various, possibly conflicting reasons. Often, this disagreement will be reasonable, in the sense just described. Most political decisions will require assessing and balancing many complicated and competing considerations. Not all reasonable people will give the same weight to these various factors, so we should expect reasonable disagreement on many, if not most, political decisions. As Rawls points out, "unanimity of views is not to be expected" (2005, 479). Still, in order for a decision to be truly collective and therefore legitimate, it must reflect reasons that all reasonable people can recognize and share, even if there is reasonable disagreement about which reasons are decisive.

Combining this requirement with the fact of reasonable pluralism generates the core of Rawls's account of public reason. The basic idea is that when a society makes a collective decision through its political institutions, it should rely on reasons that can be shared by reasonable citizens, despite the fact that they disagree about many fundamental matters. A consideration will count as a good reason for the exercise of political power only if it can be recognized as a good reason, even if not a decisive one, by all reasonable people. We assume that reasonable people, despite fundamental disagreements about their comprehensive religious, philosophical, and moral doctrines, agree with one another that in political matters, they should treat one another as free and equal citizens, capable of participating in a fair system of cooperation. When understood in this way, reasonable citizens owe to one another a justification of their political conduct in terms that they all can reasonably be expected to accept despite their reasonable disagreements. This is enough to generate robust political values and associated public purposes that public reason can recognize.

In addition to sharing a commitment to these political values, each reasonable citizen will also have values and beliefs that are not generally shared as part of a comprehensive religious, metaphysical, or moral doctrine. Such a doctrine may be fully comprehensive, if it "covers all recognized values and virtues within one rather precisely articulated system" (2005, 13), or only partially comprehensive, if it covers only some of the values and virtues beyond the political values recognized by public reason. Although these comprehensive doctrines go beyond the limits of public reason, it is crucial to understand that public reason does not deny the truth or importance of those additional beliefs and values. It simply does not engage with them, neither affirming nor denying them; it remains agnostic concerning these further matters. However, public reason holds that regardless of their truth, they do not provide good reasons for the exercise of political power precisely because we cannot expect them to be shared by all reasonable people. If one is committed to understanding society as composed of free and equal citizens, then one will have a strong basis for holding that reasons tied exclusively to a particular comprehensive doctrine, despite being true, will not be good public reasons.

Still, some people will think that the truth of their religious conviction authorizes them to use the coercive apparatus of the state to enforce their beliefs on others. Such fundamentalists are, in Rawls's technical vocabulary, unreasonable. They do not view legitimate political decisions as expressing the collective will of free and equal citizens and they do not recognize the duty to provide justifications of political decisions in terms that others can accept. Rather, they view political decisions as opportunities to advance the ends dictated by their faith, despite the fact that others reasonably reject those ends. Such individuals will often reject the political values embraced by public reason. Political liberalism does not cave in the face of such positions. It embraces and affirms a robust set of political values and holds that those who reject them are unreasonable and mistaken. We do not simply seek to accommodate whatever doctrines, whether reasonable or not, happen to exist in society at any given time. That, Rawls says, would be "political in the wrong sense" (2005, 142). So in one sense, political liberalism is most definitely a moral doctrine. It is, however, a self-limiting moral doctrine that carves out political values that are shared among all reasonable individuals from broader comprehensive doctrines about which reasonable individuals will often disagree.

A pair of examples might help to clarify the point. Suppose that Jack holds that no one should eat pork. He takes literally the prohibition in Leviticus and holds that eating pork contaminates and makes one's soul impure. Because of this faith, he holds that there is a good reason for everyone not to eat pork, whether they recognize it or not. He also knows that not everybody takes the Bible literally. He thinks such nonbelievers are mistaken. Yet he can also recognize that at least some of those who do not take the Bible literally hold reasonable, albeit, he believes, false doctrines. Perhaps if they had been raised in different circumstances, or touched by God, they too would have been believers. But despite their mistaken beliefs, Jack is committed to sharing political society with these individuals and to treating them as free and equal citizens. Therefore, he will hold that political power can properly be exercised only for reasons that all reasonable people can share. Jack can reach the following conclusion: despite the fact that the Bible provides a good reason for everyone to refrain from eating pork, it does not provide a good public reason for prohibiting its consumption through the use of state power.

Consider now Jill, who also holds that no one should eat pork. She believes that the risk of trichinosis and other diseases is prohibitive, so there is a good reason for everyone not to eat pork, whether they recognize it or not. A concern for health is not tied to any particular comprehensive doctrine. All reasonable people recognize it to be a valid concern. The risk of contracting trichinosis and other diseases counts as a public reason for prohibiting the consumption of pork. To say that there is a public reason for prohibition is not yet to say anything about the strength of that reason or whether there are countervailing reasons. In fact, we should expect that some people will judge the risks of trichinosis to be small compared to their preference for eating pork. Still, all reasonable people should be able to recognize that Jill's concern expresses a valid public reason.

These examples bring out several important points. First, a concern for health can be grounded in the political values recognized by public reason. Disease or infirmity can interfere with the ability of individuals to participate in social institutions as free and equal citizens. If, as we assume, all reasonable people are committed to ensuring the capacity of all to participate as free and equal citizens, then they will be moved by considerations of what is necessary to maintain public health. To repeat: they may often disagree about which specific policies are necessary or how the costs are to be balanced against other important political values, but they will recognize such considerations as appropriate bases for public policy. In contrast, second, a belief that certain dietary restrictions are necessary in order to maintain a pure soul carries no weight in support of a policy to enforce those restrictions. Such a concern extends beyond political values that are shared by all reasonable citizens. But public reason does not deny Jack's faith or dispute his objections to eating pork. It merely holds that such considerations, whatever their merit on comprehensive religious, metaphysical, or moral grounds, cannot provide legitimate grounds for public policy in a society of free and equal citizens. Finally, note that the idea of public reason itself does not dictate an answer to the question of whether a society should prohibit the consumption of pork on health or other public grounds. That is a decision that must be made by the citizens of a democratic society, presumably through their representatives. An account of public reason provides the framework to decide what kinds of considerations are relevant to that public deliberation and what kinds are not. At least in the case of ordinary political decisions, it is left to citizens collectively to assess the strength of these considerations and make the decisions they think best from within this framework.

Discussions of Rawls's account of public reason often present it as imposing limits on freedom of speech, but this is a mistake. It is fundamentally about what is to count as a good reason for the exercise of political power. It is therefore to be applied to what we might call public deliberations about the use of political power. In the first instance, this applies to the arguments that politicians and government officials make in their official capacities for and against various political decisions. They should invoke and consider public reasons and not rely on reasons that derive exclusively from their own comprehensive doctrines. In a democratic society, participation in political decision making is not limited to an elite group of officials but is open to all citizens. Therefore, in political deliberations addressed to the society as a whole, individuals should also invoke public reasons and not rely on reasons that derive exclusively from their own comprehensive doctrine. As a matter of free speech, individuals should be allowed to violate this requirement, just as they should be permitted to argue from insufficient evidence, commit non sequiturs, or affirm the consequent. But these are all bad arguments and should not count as supporting the use of political power. The account of public reason sets out another ideal of the kinds of reasons that should properly count as supporting the exercise of state power. Someone whose advocacy of a policy violates the requirements of public reason may be displaying a moral fault by advocating the illegitimate use of state power, but at least in most circumstances this should be a legally protected activity.

Public Reason

I need to address two additional points before turning to issues of synthetic biology. First, Rawls endorses what he calls a "wide" view of public reason. According to this interpretation, we may "introduce into political discussion at any time our comprehensive doctrine, religious or nonreligious, provided that, in due course, we give properly public reasons to support the principles and policies our comprehensive doctrine is said to support" (2005, 453). What this means, in effect, is that there is no prohibition at all on citing reasons that are based on one's particular comprehensive doctrine as long as one also cites reasons that can be shared by those with different comprehensive doctrines. Of course, even this mild requirement holds only when the exercise of political power is at stake. When it is not, there is no need to invoke the concept of public reason at all. It remains perfectly consistent with this approach to conduct vigorous public debates concerning rival comprehensive doctrines. This account provides no prohibition on individuals (or groups) engaging in theological debates, for example, as long as the exercise of state power is not at stake. Jack can certainly make his case that individuals should adopt the Bible as literal truth and therefore give up eating pork. Indeed, encouraging such debate can serve several valuable public functions. Most obviously, it can help individuals and groups clarify the nature of their religious faith and potentially change it in light of critical reflection. Political decisions are not the only things that matter, after all, and if it is important that one's religious views be correct, engaging in public theological debates may contribute to this. In addition, there can be a valuable political purpose served by introducing one's full comprehensive doctrine into public debates, as long as properly public reasons are also introduced. Although deep disagreements will continue to exist, by explicitly stating the foundational comprehensive beliefs on which one's commitment to public reason is based, one may more clearly display that one's political arguments are being made in good faith. Still, assuming that reasonable pluralism persists, considerations that can be based only on some particular comprehensive doctrine are an inappropriate basis for the exercise of political power.

Finally, I depart from Rawls on one important point in this discussion. He himself restricts his account of public reason to what he calls "constitutional essentials and matters of basic justice" (2005, 214, xlviii). When it comes to what we might call ordinary political decisions, his attitude is somewhat unclear. Sometimes he appears rather firm that the account of public reason does not apply to such cases, writing, for example, that in such cases it is "neither attainable nor desirable" to achieve "publicly based justifications" (Rawls 2001, 91n13).[3] However, he does not argue directly that the account of public reason should not apply to ordinary political deliberations. Rather, his position seems to be that, at least initially, we should focus on developing an account for constitutional essentials and matters of basic justice and then consider the extent to which it applies to more ordinary cases. Thus, he writes,

> A full account of public reason would take up these other questions and explain in more detail than I can here how they differ from constitutional essentials and questions of basic justice and why the restrictions imposed by public reason may not apply to them; or if they do, not in the same way, or so strictly . . . my aim is

to consider first the strongest case where the political questions concern the most fundamental matters. If we should not honor the limits of public reason here, it would seem we need not honor them anywhere. Should they hold here, we can then proceed to other cases. Still, I grant that it is usually highly desirable to settle political questions by invoking the values of public reason. Yet this may not always be so. (Rawls 2005, 215)

So Rawls's suggestion that the account of public reason should be applied only to the basic structure seems to be largely strategic rather than principled. He is surely right that if public reason does not represent an appropriate ideal for a society when addressing matters of basic justice, then it is unlikely to be an appropriate ideal for public deliberation concerning more ordinary political matters. Violations of the ideal of public reason when matters of basic social justice are at issue are especially offensive to the ideals of democratic citizenship. But this is not a reason why the account should not also apply to ordinary political decisions. Indeed, Rawls recognizes that "it is usually highly desirable to settle political questions by invoking the values of public reason."

Given the admitted desirability of applying public reason more broadly, why would Rawls worry that it might not be an appropriate ideal for addressing ordinary political matters? At several points, Rawls argues that it is important to establish that public reason is "complete" in the sense that it recognizes considerations that can be "balanced or combined" to "give a reasonable public answer to all, or to nearly all, questions involving the constitutional essentials and basic questions of justice" (2005, 225, cf. 454–455). In contrast, it might be thought that public reason would not have the resources to resolve more concrete disputes (Quong 2004). If that were the case, then insisting on "publicly based justifications" would be undesirable precisely because the ideal would be unobtainable. T. M. Scanlon, for example, argues that "it does not . . . seem plausible that a political conception—which must refrain from taking sides on issues on which reasonable comprehensive views may disagree—could provide the basis for answering all questions that arise in the course of legislation" (Scanlon 2003, 163). However, this is a misleading formulation of what the ideal of public reason requires. Citizens need not "refrain from taking sides on issues on which reasonable comprehensive views may disagree." What is required is that they defend their political judgments on the basis of values and ideals that all reasonable citizens can be expected to share. This does not mean that all reasonable citizens will interpret, apply, and weigh these shared values in exactly the same way in all concrete cases. On the contrary, as we have seen, we should not expect unanimity of views among all reasonable citizens: "Reasonable political conceptions of justice do not always lead to the same conclusion; nor do citizens holding the same conception always agree on particular issues" (Rawls 2005, 479). So citizens *may* take sides on issues where they expect other reasonable citizens to disagree.

Citizens must base their arguments for public policies on shared public values and ends that are not tied exclusively to their own particular comprehensive doctrine. They will, of course, need to rely on their judgment to interpret and apply these more abstract values and ideals to the concrete cases at hand.[4] The application of the abstract political values of public reason to the concrete issues of ordinary politics may follow a long interpretive path, drawing on particular and specialized knowledge. As Jonathan Quong notes, "although the content of public reason is fixed (in the sense that reasons must be mutually acceptable to reasonable people), its *detailed* content cannot be determined in advance of the process of public reasoning over particular cases in all their complexity" (Quong 2004, 244–245). This process may generate a wide range of opinions that must be recognized as reasonable, and we may reach a standoff among reasonable views concerning concrete interpretations of shared abstract values. In such a case, a society must rely on a fair political mechanism, such as the majority vote of fairly elected representatives, in order to make a decision. As long as the mechanism is just, and the views held are reasonable and good-faith interpretations of legitimate public purposes and values, the ideal of public reason is upheld.

The ideal of public reason is appropriate in the case of ordinary political decisions for the same reason it is appropriate for constitutional essentials and matters of basic social justice: justification in terms of public reason is a requirement for the legitimacy of law. In order for the coercive imposition of law to be morally acceptable, it must serve properly public ends and shared political values and ideals. Otherwise, it is simply a matter of some individuals usurping the power of the state to coerce others into serving their own ends. These ends need not be narrowly selfish in order to be politically illegitimate. Someone who advocates enforcing a prohibition on eating pork on religious grounds may not be selfish. Nonetheless, such an advocate would be failing to respect his fellow citizens as free and equal members of the shared political society. Thus, when it comes to a political decision regarding policy toward synthetic biology, whether or not it is regarded as a matter of basic social justice, debate among free and equal citizens is properly guided by the ideals of public reason.

The Politics of Synthetic Biology

Synthetic biology encompasses a wide range of emerging technologies that have elicited a variety of arguments for and against. I want to underscore once again that I am here concerned only with the basis on which political decisions to regulate, to support, or to prohibit these technologies ought to be based. Nothing I say should be understood as rejecting or criticizing the consideration of ethical implications from the point of view of some particular comprehensive doctrine, whether religious or secular. When the use of political power is at stake, however, we should aim to identify what are the strongest public reasons. Public purposes, recognized by public reason, are clearly implicated in the development of synthetic biology. We'll discuss some of these below. But certain considerations and arguments fall outside of its scope. Specifically, many arguments that express a so-called intrinsic objection fall afoul of public reason.

It is important to note that the problem with the intrinsic objection to synthetic biology is not that it invokes intrinsic values. Public reason does so as well, recognizing justice itself and the political values that derive from it as intrinsically valuable. The problem is that the intrinsic objection typically does not invoke public political values. So while the values and moral principles on which such arguments are based may or may not be sound, when they are based on a particular comprehensive doctrine (or range of doctrines), they are not appropriate bases for making a political decision.

For example, writing in 1997, following the cloning of Dolly the sheep, Leon Kass claimed that although "revulsion is not an argument," it is—or can be—"the emotional expression of deep wisdom, beyond reason's power fully to articulate it" (Kass 2000, 79). Kass took his own feelings of repugnance, and the purported feelings of repugnance among the majority of citizens, to constitute a reason to ban human cloning. Despite his confident assertion, not all reasonable people feel a deep disgust at the thought of human cloning. Kass is free, of course, to make his case that they should. But regardless of whether they do or should, repugnance, even if felt by a majority, by itself is not a public reason. It should not count as a reason for public policy at all.

Similarly, an argument based on anything like a direct appeal to a religious text or tradition falls outside of the limits of public reason. So, for example, an argument purporting to show that synthetic biology is contrary to Biblical teachings falls outside of public reason. Someone who would use political power to prohibit it on this basis alone would be unreasonable in her attitude toward her fellow citizens. She would be treating them not as coauthors of their society's political decisions, entitled to reasons that they can recognize in their own terms, but merely as tools or instruments for pursuing the more important ideals and goals that she finds in the Bible. It is often said, in criticism of various forms of genetic intervention, that we should not "play God." Many different objections are expressed in this way, but taken in its most direct and literal sense, the criticism amounts to the claim that some activity is blasphemous—that it "confuses our role with God's" (Sandel 2007, 85). And this, no less than the other examples we have seen, whether true or not, does not provide grounds for the use of political power in a just democracy.

There are, however, secular interpretations of this objection to consider. The worry expressed in the phrase "playing God" may not literally be that our conduct would offend God, but rather that it is an objectionable expression of hubris. This seems to be the view offered by Michael Sandel. While Sandel defends the use of genetic interventions for medical purposes—for "restoring and preserving the natural human functions that constitute health" (2007, 47)—he is opposed to interventions for the purpose of enhancement. I want to accept for the sake of argument the distinction between treatment and enhancement, and consider some of the arguments that Sandel gives against genetic intervention for enhancement. Some of the reasons that he gives fall outside the scope of public reason, but others can be recognized within it. Ultimately, it seems, Sandel's objection is to the attitude that he believes would be expressed and reinforced by parents who would genetically enhance their children. In fact, Sandel finds this attitude quite prevalent among parents even absent genetic modifications. "These days," he tells us, "overly ambitious parents are prone to get carried away with transforming love—promoting and demanding all manner of accomplishments from their children, seeking perfection" (2007, 50). There is currently an "epidemic of parental intrusiveness and competitiveness" (p. 53), and a "frenzied drive by parents to mold and manage their children's academic careers" (p. 54). This suggests that his concern is not intrinsic to genetic manipulation, and when pressed, he concedes the point: "Some see a bright line between genetic enhancement and other ways that people seek improvement in their children and themselves. . . . But morally speaking, the difference is less significant than it seems" (p. 61). The claim, in other words, is that genetic enhancement (but not it alone) threatens to "transform three key features of our moral landscape—humility, responsibility, and solidarity" (p. 86).

These values can be endorsed, at least in some form, from within public reason. They are not tied to only one or a few comprehensive doctrines. Therefore, we can imagine a form of Sandel's argument against genetic enhancement being presented within public reason.[5]

Notice, however, three features of Sandel's argument. First, as we saw, it is not an intrinsic objection to enhancement. Rather, the claim is that such interventions are objectionable because they are expressions of and contribute to an attitude toward children that is objectionable. If it could be shown that the desire for enhancement was motivated by a different attitude or that it did not tend to undermine humility, responsibility, or solidarity, then Sandel's objections would disappear. Second, it is crucial to his argument that his objections are to enhancement, not to medical interventions that are guided "by the norm of restoring and preserving the natural human functions that constitute health" (2007, 47). Third, since the allegedly objectionable attitude is toward children, there does not seem to be any analogous objection to the genetic manipulation of crops, for example, or more broadly to synthetic biology. So, while Sandel offers his argument as a secular analog to the religious concern with "playing God" (2007, 85–86), in fact his argument does not even purport to show that genetic manipulation is intrinsically wrong. Although some of Sandel's arguments can be interpreted as presenting public reasons to prohibit genetic enhancement, this does not mean that they are strong arguments or that there are not countervailing public reasons. One particular weakness in his argument is the fact that he discounts any possible motivation for genetic manipulation other than the "drive to master the mystery of birth" (2007, 46).

Synthetic biology also raises other concerns that can be expressed in terms of political values. The most obvious of these relates to environmental issues. Some forms of environmentalism go beyond political values and represent a comprehensive doctrine, expressing views that are not shared by all reasonable people about the ultimate moral standing of the natural world or the metaphysical relationship between humans and nature. For example, if deep ecology holds that the environment itself has a right to life on a par with that of humans, this is obviously a doctrine that goes well beyond shared political values. Yet all reasonable comprehensive doctrines will go as far as to endorse some form of environmental conservation. Each, perhaps for its own foundational reasons, will recognize that a serious threat to the environment is a threat to everyone's ability to live as free and equal citizens according to their own reasonable doctrine with its system of ends and values. If synthetic biology threatens significant environmental harm, this certainly can be the basis for a public reason to regulate or even prohibit certain risky activities.

Another political value that may be at issue in discussions of synthetic biology is the idea of equal opportunity. This value is not dependent on any particular comprehensive doctrine. As it has been interpreted in the literature, it has two components. First, in developing his principles of justice, Rawls assumes that the individuals for whom the principles of justice are selected are "normal and fully cooperating members of society" (Rawls 2001, 171–172). This obviously is an idealizing assumption, and any actual society needs to develop principles of justice to deal with individuals who either permanently or temporarily fall below this threshold. Norman Daniels has developed an account of just health care under the rubric of Rawlsian fair equality of opportunity, where the goal is to restore normal human functioning.[6] Rawls has followed Daniels on this point, holding that fair equality of opportunity requires that we recognize claims of medical care to restore "persons to good health, enabling them to resume their normal lives as cooperating members of society" (Rawls 2001, 174). However, the threshold of normal human functioning required to be a fully cooperating member of society can change as society and technologies develop. As Allen Buchanan and his coauthors argue, "If it becomes within our power to prevent what we now regard as the misfortune of a sickly constitution (a weak immune system) or the catastrophe (the natural disaster) of a degenerative disease such as Alzheimer's dementia, then we may no longer be able to regard it as a misfortune. Instead, we may come to view the person who suffers these disabilities as a victim of injustice. As our power increases, the territory of the natural is annexed to the social realm, and the new-won territory is colonized by ideas of justice" (Buchanan et al. 2000, 83–84).

The other component of fair equality of opportunity assumes that everyone has normal human functioning above that threshold, but that there is, nonetheless, variation above that level. In that case, Rawls specifies the ideal of fair equality of opportunity as follows: "Assuming that there is a distribution of natural assets, those who are at the same level of talent and ability, and have the same willingness to use them, should have the same prospects of success regardless of their initial place in the social system" (Rawls 1999, 63). Rawls is especially sensitive to how the design of the social system influences the development and expression of natural talents and capacities. Still, the specification of the ideal of fair equality of opportunity depends on a distinction between natural and acquired talents. This line is and will become even more blurry. We will have to rethink the implications of this ideal. Without exaggerating the extent of our control, it is fair to say that as social control encompasses what was previously subject to the natural lottery, the demands of fair equality of opportunity will expand. So the two components of equal opportunity can be specified as follows: "equal opportunity has to do with ensuring fair competition for those who are able to compete *and* with preventing or curing disease that hinders people from developing the abilities that would allow them to compete" (Buchanan et al. 2000, 74). Both of these components, recognized within the limits of public reason, will be affected by developments in biotechnology, and both, I suggest, will lead justice to make demands that are more egalitarian. Some of these demands may have direct implications for the funding or regulation of the technologies of synthetic biology.

Just as some arguments for prohibition exceed the limits of public reason, it is also not hard to imagine arguments supporting synthetic biology that exceed those limits. If someone were to argue that we should create new forms of life as a way of completing God's purpose for us or in order to fulfill our destiny as masters of nature, these arguments would obviously rely on controversial comprehensive values that some reasonable people would reject. I have not dwelled on such arguments because there are much more straightforward arguments for synthetic biology that are accessible from within the framework of public reason. The potential practical benefits promised from synthetic biology are immense, including new drugs and medical treatments, clean energy sources, and other environmentally friendly products. The development of these and many other applications can be recognized as valuable from within public reason. Their justification can be tied to shared public purposes rather than to any particular comprehensive doctrine.

Although the intrinsic objection to synthetic biology is not a good basis for legitimate policy, there remain powerful public reasons against its uncritical pursuit. It is simply a mistake to assume that intrinsic objections are always more powerful than what we might call extrinsic objections. As indicated above, the most obvious objections recognized within public reason are related to various kinds of environmental risks. We are only now coming to appreciate the fragility and complexity of environmental ecosystems. Public reason can recognize environmental and social considerations both for and against synthetic biology generally, just as it can recognize considerations both for and against the use of genetically modified crops. My reading of our environmental record with respect to genetic modification of crops is that it is mixed, but our treatment of another fragile environmental system threatens to be catastrophic. Our apparent inability to cope with global climate change tells in favor of caution in exploring and deploying new technologies, the environmental consequences of which are so unpredictable. But the framework of public reason is only that—a framework within which a democratic society assesses the reasons for and against various policy decisions.

Those who cite intrinsic objections to synthetic biology or other genetic intervention as the basis for public policy exceed the limits of public reason and seek to deploy political power in an illegitimate way. Still, they are not unreasonable simply because they hold that synthetic biology is intrinsically objectionable. While public policy should not be based on intrinsic objections, it should not be based on the assumption that those who argue that that synthetic biology is an affront to God are necessarily mistaken or unreasonable. If one affirms a comprehensive doctrine that rejects synthetic biology without attempting to enforce this view through the use of state power, one is not being unreasonable. And if a society makes a legitimate decision to pursue the use of a new technology that some reasonably reject, society should make every effort to accommodate those who reasonably reject it. In the case of genetic modification of crops, one obvious response is a policy of mandatory labeling (Streiffer and Hedemann 2005). Accommodation in other areas may be more difficult. For example, patient autonomy is undoubtedly an important consideration in deploying any medical technology, including new genetic interventions. Decisions to forgo such interventions should often be respected. However, there is a significant risk that those who refuse such treatments may be marginalized in various ways. Such individuals are still owed fair equality of opportunity and therefore reasonable accommodation. It may be difficult to avoid various stigmas, but that is what is required in circumstances of reasonable pluralism.

I conclude by emphasizing one final time that the reasons that political liberalism gives for respecting the limits of public reason are themselves moral reasons. Political liberalism is not based on an aversion to disagreement as such or on skepticism about the possibility of moral truth. Political liberalism is a fighting doctrine. Grounded in an account of the requirements for the legitimate use of political power, it holds that those who reject it are unreasonable and mistaken. These are moral considerations, but because of the conditions of reasonable pluralism in modern free societies, these moral reasons themselves restrict the class of other reasons to which we may appeal in public deliberations over the use of state power. In many ways, this picture is simply an extension of the ideals of liberty of conscience and of the toleration of diverse religious faith. Just as justice requires that we tolerate religious faiths that we believe to be mistaken, so too it requires that we offer each other reasons for the deployment of political power that we expect that others can reasonably accept.

Acknowledgments

A version of this paper was presented at The Hastings Center on March 16, 2010. I am grateful to the audience there and to Pete Murray for discussion and suggestions.

Notes

1. There Rawls notes that "it is possible to adopt eugenic policies, more or less explicit. [However,] I shall not consider questions of eugenics, confining myself throughout to the traditional concerns of social justice." He goes on to mention that it makes sense for society "to take steps at least to preserve the general level of natural abilities and to prevent the diffusion of serious defects" (p. 92).
2. This edition includes "The Idea of Public Reason Revisited," from which I will be quoting below.
3. Cf. Rawls 2005, 235, 246.
4. Rawls notes that "There are also different interpretations of the same conception [of justice], since its concepts and values may be taken in different ways. There is not, then, a sharp line between where a political conception ends and its interpretation begins, nor need there be. All the same, a conception greatly limits its possible interpretations; otherwise discussion and argument could not proceed" (2005, 455n35).
5. I should note that Sandel himself is not concerned to present his argument in terms acceptable to public reason.
6. See, for example, *Just Health Care* (Cambridge University Press, 1985), and *Just Health: Meeting Health Needs Fairly* (Cambridge University Press, 2007).

References

Buchanan, Allen, Dan Brock, Norman Daniels, and Daniel Wikler. 2000. *From Chance to Choice: Genetics and Justice*. Cambridge, UK: Cambridge University Press.

Kass, Leon. 2000. The wisdom of repugnance: Why we should ban the cloning of humans. Reprinted in *The Human Cloning Debate*, ed. Glenn McGee. Berkeley, CA: Berkeley Hills Books.

Quong, Jonathan. 2004. The scope of public reason. *Political Studies* 52 (2):233–250.

Rawls, John. 1999. *Theory of Justice*. Revised edition. Cambridge, MA: Harvard University Press.

Rawls, John. 2001. *Justice as Fairness: A Restatement*. Ed. Erin Kelly. Cambridge, MA: Harvard University Press.

Rawls, John. 2005. *Political Liberalism*. Expanded edition. New York: Columbia University Press.

Sandel, Michael. 2007. *The Case against Perfection: Ethics in the Age of Genetic Engineering*. Cambridge, MA: Harvard.

Scanlon, T. M. 2003. Rawls on justification. In *The Cambridge Companion to Rawls*, ed. Samuel Freeman. Cambridge, UK: Cambridge University Press.

Streiffer, Robert, and Thomas Hedemann. 2005. The political import of intrinsic objections to genetically engineered food. *Journal of Agricultural and Environmental Ethics* 18 (2):191–210.

8

Biotechnology as Cultural Meaning: Reflections on the Moral Reception of Synthetic Biology

Bruce Jennings

Perhaps the fundamental question before us in science policy today involves the extension of human power and artifice into the realm of life. The general question is not new. Shakespeare's Prospero pondered it, as did Mary Shelley's Frankenstein and H. G. Wells's Dr. Moreau. But the gap between fantasy and actual technological capacity is closing, so that now the morality of power must speak to the governance of power; ethics must inform public policy. Synthetic biology constitutes a significant extension of the human capacity to manipulate the conditions of life at several levels—the molecular and cellular level, the level of organ systems and metabolic pathways, the traits and behavior of individual organisms, and the level of ecosystemic interactions and interdependencies.

The advance of human capacity to manipulate the conditions of life is generally referred to as biotechnology, and the extension of the domain of human agency that it brings about may be referred to as "biopower" (Rose 2006). Whenever there is an innovative, substantial, and rapid extension of human power—as is happening with synthetic biology—society must come to grips not only with the scientific and political implications of that power but also with its cultural meaning. We try to make rules to protect ourselves, and we try to fashion a story about this new kind of power that will domesticate and civilize it. This is necessary because in its raw or wild form such power is too exhilarating, too frightening, too dangerous, and too open to abuse to coexist comfortably with our settled ways of life.

In the United States, if not in Europe, thus far the narrative framing of biotechnology is largely positive, except perhaps within the sensitive and highly charged symbolic and semantic domain of human reproductive medicine and its next-door neighbor, embryonic stem cell research. And yet, despite its aura of technological progress and advance, an unease surrounds biotechnology generally, and I think synthetic biology throws that unease into sharp relief. There is something uncanny about fabricating life with nearly the same facility that inorganic matter and energy are manipulated. I believe that this unease does not stem solely from the concern that biotechnology will be misused by human agents. It also grows out of the realization that institutionalized structures of power (state or corporate) have an agency of their own, so that power is not something we use or abuse, *it is something that uses—or abuses—us*.

The advent of synthetic biology therefore requires consideration from a political and cultural perspective and not merely a scientific or technical one. In this chapter I propose to focus on the cultural and social meanings of synthetic biology, which are various and highly contested, and on the ethical implications of this exercise in cultural hermeneutics. This kind of interpretive inquiry is important because it plays a key, albeit oftentimes overlooked, role in the formation of science policy in a democracy; and it does so because democratic governance is based not solely on expert knowledge and opinion but also on a broader kind of public perception and legitimacy.

Biotechnology and Culture

In order to grasp how synthetic biology will be received and reacted to in the political culture of the United States today, we should concentrate not on what is most specific and distinctive about it biologically and technologically, but instead on the broader significance, interpretations, and concerns that are being imputed to biotechnology writ large (Evans 2002). The semantic and symbolic framing of synthetic biology and its political narrative will be based on synthetic biology not as a unique *type* of research and technology, but more as a *token* of a larger technological revolution, with all the confusion, hope, and anxiety that the revolution is now stirring in our moral and social imagination. Moreover, no technology (including a biotechnology like synthetic biology) should be defined it terms of its apparatus and instrumentality alone. Beyond machinery or physical apparatus, technology is a complex system incorporating both engineering and sociological elements. In its social aspect, a technology is a specific formation of human organization, institutional power, and symbolic meaning.

I wish that these remarks and this orientation were a commonplace, but in fact they go against the grain of many influential approaches to social theory and technology studies. In the mid-twentieth century, during the heyday of scientific American social science, many theorists, such as the noted anthropologist Leslie White, developed what was sometimes referred to as the "layer cake" model of technology studies, in which the meeting of material needs via technology was at the base, social institutions and rules were the middle layer, and symbolic meaning was, as it were, the top layer and icing (Sahlins 2010). Again, the upshot of this tradition, in what has been called the "first wave" of science and technology studies, was to understand culture as a projection of underlying material conditions, including the physical (or biological) requirements of the economically dominant technology of a given society, in a given environmental setting, at a given time (Collins and Evans 2002).

On the other hand, more recently various schools of thought in the social sciences have stressed the role of symbolic formations in a given culture—normative discourse, intellectual traditions, metaphysical or ontological beliefs, the use of metaphor and other kinds of figurative and narrative reasoning both in lay understanding and in the process of scientific discovery and creativity, and the like. These symbolic formations shape technology and the material conditions of the production and reproduction of human social life as much as they are shaped by those conditions. This reciprocal interplay between material conditions and cultural meaning—what the social theorist Anthony Giddens (1976) refers to as the double hermeneutic—sheds an interesting light on the historical question of which specific technologies are discovered (and even conceivable as such) at particular times, and which are not. Societies never develop all the technologies or material possibilities open to them in a given time and place; the actual array of technologies discovered, invented, and deployed is always a subset of the possible array. There is no reason to think that this selection is merely accidental or fortuitous.

I am inclined to side with the culturalists against the materialists in this ongoing debate within history and social science. The social interpretation and meaning of biotechnological innovation are often overlooked in policy analysis, but they are key to a deliberative process of social learning and adjustment, and to the normative consensus formation that will allow any regulation of biotechnology to be truly effective. The policy process, as is well known, is often attuned quite differently from the cultural and political world, and the dissonance between these worlds often frustrates and confounds policy elites and technical experts, who seek propositional assertions and linear reasoning in what is more fundamentally a narrative and figurative mode of discourse. Ethics, as I understand it, regularly must assume the rather awkward posture of straddling these distinct modes of discourse and their practical applications.

Moreover, there is a dimension to the governance of biotechnology, the taming of biopower, that mainstream policy thinking regularly overlooks, but which is well established in our philosophical, ethical, and religious traditions. Policy analysis has a tendency to see power and technology as neutral tools for the extension of human agency and intentional use. But from other perspectives, which I believe are widespread and influential in American political culture today, power must be tamed and governed, not only—or even primarily—because it will be misused by human agents, but because, as I noted at the outset, institutionalized structures of power have an agency of their own that stands over above the will and control of the individual human agent.

The Meaning of Biotechnology

What is the best way to organize an exploration of the semantic framing and the political narrative of biotechnology? The first step is to place the cultural reception of biotechnology and synthetic biology in a historical and ideological context. Following that, I propose to discuss some of the salient types of questions or concerns that guide the development of these frames and narratives.

I wish to single out four broad factors in the recent historical and ideological context affecting the cultural reception of biotechnology. The first is the return and politicization of value questions in science. The second is the sense of loss of self-determination. Third is the sense of the loss of self-control, which is not the same thing as self-determination. Fourth is the loss of trust in the functioning and the integrity of major social institutions.

The End of Value Neutrality and the End of Liberal Optimism Concerning Science and Technology

In the postwar period, a consensus gradually developed, especially in the United States and perhaps somewhat less enthusiastically in Europe, concerning authority, expertise, and progress in science and technology. It was a consensus centered around progressive values, economic growth, social modernization, and the betterment of life through technological advance. This consensus emerged during a buoyant time in the political culture of liberalism, a time of optimism about the ability of the social sciences to inform public policy, a time of enthusiasm for endeavors such as urban planning and social engineering. Speaking to the White House Economic Conference in 1962, President John F. Kennedy articulated a then widely held view of the challenge facing government in the United States:

> The fact of the matter is that most of the problems, or at least many of them that we now face, are technical problems, are administrative problems. They are very sophisticated judgments which do not lend themselves to the great sort of "passionate movements" which have stirred this country so often in the past. Now they deal with questions which are beyond the comprehension of most men. (Kennedy 1962)

Less than a decade after President Kennedy declared their irrelevancy, passionate movements returned to the domain of public policy with a vengeance and shattered the notion that significant social policy areas, such as foreign policy, economics, health, the environment, and science, were merely "technical problems."

For social and cultural conservatives on what is often referred to as the religious right, the framing of the right-to-life movement—often based on a theology of the dignity of the human person—has clearly grown beyond the legalization of abortion to encompass much of the new biotechnology, at least insofar as it touches in any way on human reproduction or germline genetic or cellular substances. In addition, controversies have flared up over the cultural meaning of evolutionary biology and the control of basic science education.

Yet critiques of biotechnology and biopower are not confined to the cultural right. On the cultural and intellectual left, there has been a turning away from the scientific rationalism and progressivism that not only informed the tone of President Kennedy's remarks in the early 1960s but have been integral to the liberal philosophical tradition for over two centuries. Thus John Stuart Mill spoke for the whole of liberalism when he saw in nature and in the frailty of the human body sources of suffering and limitations to be overcome by science and intellect. Concerns about being "unnatural" or "playing God" have never had much ethical bite among philosophers and have rung hollow on liberal ears (Mill 1985). However, a new kind of cultural and political framing has emerged in recent years that sees in the control and the overcoming of nature not the amelioration of the human condition, but the development of new forms of power and totalitarian control over individuals as material bodies—reproducing, laboring, neurochemically behaving bodies. This frame narrates the advance of biotechnology not in terms of the nineteenth-century liberal notion of progress, but in terms of the much, much darker twentieth-century political regression into totalitarian dehumanization of racism, genocide, eugenics, euthanasia, and mind control.

Anxiety Concerning Normative Chaos in the External Landscapes of Our Lives

Just as there has been a loss of confidence in the notion that science and technology equate with progress in human health and freedom, so too there has been a parallel loss of confidence in the fundamental elements of what was once referred to, without any sense of irony whatsoever, as "the American dream." The dream has faded as its basic components have receded in people's perceptions of the conditions of their own lives and of the social condition overall. These components include fairness, meritocracy, unlimited economic growth, and future betterment in terms of upward social mobility, educational and employment opportunity, and a secure and dignified old age. Perhaps most central of all is the work ethic, which is not simply a legacy of a Protestant or Puritan outlook in which toil was a way of disciplining naturally sinful human beings, but is also, as Max Weber discerned, a notion about the social conditions under which an individual could control, at least to some degree, the conditions of his or her own life (Weber [1905] 1958).[1]

Today the global economy is shifting perhaps as drastically as it did at the dawn of the capitalist era. It has become an enormous mechanism for churning individual lives. Marx once remarked that capitalism evaporated solid traditions and social relationships—all that is solid melts into air. Updating the simile, Zygmunt Bauman refers to the "liquid" nature of our society: everything—from electrons to day laborers—is fluid, fungible, flowing (Bauman 2000). Each of us is replaceable and displaceable. Even in the most affluent societies and the most powerful nation-states, individuals today are confronted with institutional experiences and cultural images of social and personal dislocation that challenge, to put it mildly, their sense of personal efficacy and control.

These experiences ultimately grow out of structural forces over which individuals as such have little or no control. These forces include global competition and the global mobility and manipulation of finance; the transition from mass-production industrial economy to electronic, service, and finance driven economy; the widening disparity and growing concentration of wealth and power in society and its political manifestation in the media attention to scandal, cover-up, and institutional abuse. Ordinary adults today cannot be secure in their employment no matter what level of effort or even productivity they achieve. They cannot control what they ingest or what health hazards they are exposed to; they cannot control the images their children see. They cannot rely on the promise of pensions and have witnessed the value of their investments and equity cut in half on the basis of reckless decisions by literally a handful of profit-seeking individuals and investment banks. They perceive the precariousness of lives that cannot avoid economic ruin if something should arise to disrupt their household income stream. Does biotechnology liberate us from this structure, or does it mesh into this brave new world all too seamlessly?

Anxiety Concerning the Moral Disorder of the Internal Landscapes of Our Selves

If the work ethic lies at the heart of our market society and capitalistic culture, so does the profit motive, the notion that "Greed works." Overall, human progress and well-being comes, not from the taming, civilizing, or repression of desire, but from its creative channeling via the impersonal, amoral mechanism of the market. The market today is becoming a God that has failed in the eyes of an increasingly large number of disillusioned believers. The truly interesting aspect, however, is not market failure, even on a global scale, but what that failure portends. Is it a cyclical feature, an inevitable adjustment, a temporary patch of poor oversight and government regulation? Or is it that the very successes and excesses of the market during the last fifty years have deeply transformed and corrupted society and the personality structures engendered by the society? Historians a generation ago, such as Christopher Lasch and Daniel Bell, warned of this phenomenon (Lasch 1978; Bell 1976). Have we now created a culture of narcissism and desire so insistent that it cannot be regulated, channeled, or constrained?

As this concern matures and spreads throughout the culture, it does so not because there is clear and convincing evidence of it, but because it is a framework that helps people make sense of what is happening to them and of what they perceive going on around them in the community and the nation. And it presents biotechnology with a significant, three-pronged challenge. First, biotechnology is being developed and marketed by private corporations that we can no longer trust to be publicly responsible or even constrained by market competition. Next, the biotechnology industry is being regulated by government agencies that are captured by the interests they regulate and whose ideological heart is not in the activity of regulation. Finally, for their part, the consuming public has no moral compass to serve as a restraining force on biotechnology, for they cannot say no—either as citizens, or as private consumers—to anything that offers them or their families health, enhancement, or longevity. If industry, government, and private consumers in the marketplace cannot regulate biotechnology responsibly, to whom should we turn? Who can be trusted?

Loss of Trust

There has been an erosion of trust in the functioning of government at all levels and across all functions—loss of trust in executive branch regulation, in a fairly representative legislative process, and in an objective judiciary; and skepticism concerning the social responsibility of corporate and private-sector institutions, academic experts, and professionals, including scientists (Fukuyama 1995).

However, behind specific forms of skepticism, disappointing policy failures, and occasional scandals, a broader ideological shift seems to be occurring, which, like the loss of optimistic faith in technological progress mentioned above, again involves a crack in the foundation of liberal reason itself. For some time much of our political culture has rested on a principled, progressive liberalism of reason, mutual cooperation, and fairness. This way of thinking was exemplified a century ago in the writings of John Stuart Mill and in our own time by John Rawls, Mill's worthy successor in the liberal tradition. Liberal reason holds out the promise of the preservation and promotion of the dignity, respect, justice, and fair equality of opportunity of all. It presides over a well-ordered society structured by just institutions; a society worthy of consensual cooperation and support by free and equal persons. Insofar as this framework provides us with a vibrant source of our ideals, values, and expectations (and I believe that it does and should), there is a growing and exceedingly important perceived gap between these standards and the actual facts of contemporary governance, policy, and practice. This gap between values and the use of power—this gap between what we believe we should be and what we actually are as a society—is creating a widespread form of cognitive dissonance in America today. The perceived failure of principled liberalism and reason to prevail over the dynamics of neoliberalism (an ideology in which market competition and success are the only normative standard for economic and political life) and the forces of corporate power has led to a breakdown of popular trust in social elites across the board, and this includes scientists. The loss of trust has been exacerbated by extreme partisanship and paralysis of government at the federal and state level, and egged on by the fragmentation of the sources of information and ideological interpretation in the media. We are simply unable to answer the age-old question—"Who guards the guardians; who watches the watchers?"

These contexts present a turbulent and inhospitable cultural climate for the arrival of a major new technological innovation (some even say a major new technological age) like synthetic biology. But, for good or ill, this is the petri dish, so to speak, within which biotechnology has been born, has developed, and is continuing to grow apace. Certain aspects of biotechnology, such as embryonic stem cell research in the United States and genetically modified food in Europe, have been focal points for controversy, while arguably more important economic, ecological, and scientific innovations have been carried on virtually under the radar of the general political culture. But no aspect of biotechnology is privileged or immune from this maelstrom of contested meanings.

Some scientists persist in the belief that if only policy makers and the attentive portions of the public understood the scientific facts and the reasonable scientific risks and benefits accurately, then the controversies over biotechnology would largely fade away, because after all they are grounded in misinformation, misunderstanding, superstition, and irrational animus in the first place. The phenomena I have been considering thus far suggest that this belief is naive. Controversy arises as much out of the lived experience of people in different locations within our economy, society, and culture as it does from scientific ignorance. And biotechnology is controversial not just among those who consider the practice of human molecular manipulation of life to be intrinsically wrong, but also—and I would even say primarily—among those who actually support its promise, the values it could serve, and the means that it uses, but who doubt that it will be controlled and governed properly so that those ends are equitably achieved. My notion is that the main kind of friction that biotechnology and synthetic biology will encounter in the future will come not so much from principled, ideological, or doctrinal rejection of the enterprise, root and branch, but from a narrative of mistrust and what has been called a "hermeneutic of suspicion" (Ricoeur 1970).

Counternarratives

I now turn from the general and inchoate cultural background confronting biotechnology to more systematic and conceptually well-developed modes of critical response. These offer counternarratives to biotechnology, or at any rate to some of its recent forms of self-presentation or deployment. I call these the precautionary frame, the liberal humanist frame, and the ontological frame. Stated differently, these are the argument from prudence, the argument from dignity, and the argument from nature. These modes of response have had relatively little traction in the policy debates over biotechnology regulation in the United States. (They have been more salient in discourse in the European Union.) Nonetheless, they are worthy of serious attention.

The Precautionary Frame, or the Argument from Prudence

This framework presents an attempt to break out of the logic of cost-benefit and risk-benefit analysis that has developed in mainstream economics and policy studies as a rational mode of decision making under conditions of uncertainty (Raffensperger and Tickner 1999). The gist of the "precautionary" critique is that this type of analysis is biased in favor of innovation and short-term benefits, while unduly discounting long-term and emergent systemic risks. The remedy, often referred to as the precautionary principle or precautionary policy making, is to place the onus on the demonstration of beneficial outcome rather than on the demonstration of harmful outcome. Note that the concept of "harm" is used broadly in this precautionary discourse. It includes biological risks and harms to functioning ecosystems, biodiversity, and human health, but it also encompasses a damage or loss to cultural resources and meaningful ways of life. A clear and distinct way of speaking of these kinds of risk or harm—the physical, material, or biologic and the cultural, the meaningful, the symbolic—is something that all of these four modes of critique wrestle with, but none has yet found.

This orientation also raises questions about the capacity of insiders, self-interested experts, or peer review mechanisms to judge whether this reversed burden of proof has been met. Accompanying the precautionary approach, therefore, is often a call for more public transparency, engagement, and participation in the governance process, so that at least some of the necessary monitoring can be performed by the skeptical gaze of outsiders and disinterested parties. (One does not have to agree with the precautionary approach to favor some more transparent and broader forms of public engagement in science and technology governance, of course.)

The Liberal Humanist Frame, or the Argument from Dignity

The focus of the liberal humanist frame is centered mainly on human dignity and on the preservation of the precious and hard-won orientation that makes the individual human person the subject of ethical concern and political rights. One part of the liberal humanist critique has centered around ecological and health risks, the malicious and weaponized use of biotechnology, and the problem of distributive justice in the access to beneficial biotechnological resources, products, and drugs. Another dimension of the liberal humanist critique is perhaps more conceptually radical. Its doubts concerning biotechnology revolve around the cultural and semantic dangers or risks of the technology and the reductionism of the science upon which the technology is based. This concern involves the prospect that biotechnology will undermine the conceptual foundations for our traditional notions of the moral worth of the human individual, human dignity, and equal rights to liberty, respect, and justice (Kass 2002; Habermas 2003; President's Council on Bioethics 2002; Sandel 2007; Fukuyama 2002; Kass and Wilson 1998; Lewis [1947] 1955; Mitchell et al. 2007; Nussbaum and Sunstein 1998). This line of argument is also deeply concerned about the cultural effects of institutionalizing and using the technology in ways that might undermine individual freedom and dignity. The latter critique has been developed in relation to such biotechnology applications as cloning (Habermas) and enhancement of normal human traits and abilities (Sandel).

What if, these critics suggest, an application of biotechnology does not so much harm human interests as undercut the rationale for saying that human interests—individual human interests at any rate—are things of moral value and significance? The liberal humanist critique concludes that ethical governance of science and technology must have standards of rightness and wrongness as well as standards of benefit and harm. Furthermore, ethics should challenge the largely materialistic conception of moral goodness and benefit that has been promulgated by modern science, or at any rate, by the cultural identity, the public image, of modern science and technology. A thicker and more robust conception of the moral good and moral possibility is needed to inform the governance of science.

Here, debates within bioethics and the ethics of biotechnology have revealed tensions that are interesting and important for contemporary political theory more generally. Earlier I mentioned the growing cultural skepticism about the traditional liberal notion of progress and about the optimistic faith in science and technology evinced in what can be called the "progressive liberalism" of the nineteenth and twentieth centuries. In the debates over biotechnology, this progressive liberalism comes into conflict with liberal humanism. The moral vision about the betterment of the human condition through the growth and application of science, and of human control based on scientific understanding, articulated by Mill and by Rawls also finds acute expression in the following formulation by Ronald Dworkin, who finds in biotechnology not a threat to the foundations of liberal individualism but a fulfillment of our historic, even evolutionary duty to preserve it:

> There is nothing in itself wrong with the detached ambition to make the lives of future generations of human beings longer and more full of talent and hence achievement. . . . On the contrary: if playing God means struggling to improve our species, bringing into our conscious designs a resolution to improve what God deliberately or nature blindly has evolved over eons, then the first principle of ethical individualism commands that struggle, and its second principle forbids, in the absence of positive evidence of danger, hobbling the scientists and doctors who volunteer to lead it. (Dworkin 2000, 452)

Note a key phrase in Dworkin's argument above, "improve our species." From the point of view of the nature-centered or biocentric semantic frame discussed below, both progressive liberals like Dworkin and liberal humanists like Sandel, Habermas, and Kass all are limited by the human-centered or anthropocentric perspective they adopt. Here we touch on an issue with a detailed historical past and multifaceted contemporary significance. Again, Dworkin is clear and instructive on the logic of the anthropocentric vision. He argues that there is nothing wrong with focusing human design and action on improving a creation that God deliberately left imperfect or that nature blindly evolved.

So the theocentric and the biocentric (nature transcendent and nature alive) are set aside. Is biotechnology all about the human? Does nature, or creation, or what once was called cosmos really have nothing to do with it? The third framework of cultural meaning and critique that I want to discuss answers these questions with a resounding, no.

The Ontological Frames, or the Argument from Nature

Ethical analysis is deeply affected by the ontological starting point or orientation one assumes. The cultural reception of science is deeply affected by ontological orientations that pose the question of the "right relationship" between human agency and the rest of being. To answer this question, natural being (nature or the natural) must be conceptualized. In general, there are three concepts of nature that are germane to our topic. I would call these nature dead, nature transcendent, and nature alive.

Nature dead holds that the difference between human being and all other forms of natural being is a qualitative difference (a difference in kind and not merely in degree), and it is anthropocentric in the sense that it privileges human being over all other (inferior) forms of natural being. Only human life truly matters; only human beings are alive in a metaphysically substantive sense. According to this ontology or worldview, nature, or natural being, is material without meaning, except insofar as it serves human purpose. This is the ontological frame most often used to defend and promote advancing biotechnology and augmenting biopower. Biotechnological engineering is "natural" because nature is simply raw material to be "improved" by human intelligence in the service of human well-being.

Nature transcendent places all being in a teleological narrative—being in a becoming toward fulfillment. The norm of right relationship for humanity is to accommodate and live in accordance with that narrative. When that narrative is thought to have a transcendent author, a divine Creator, the Being of all being, this ontological frame may be said to be religious, as well as philosophical and ethical. This ontology assesses human activity in the context of conceptions of the purpose for which human beings exist, the transcendent (divine) plan for human existence, and the proper relationship between human and nonhuman modes of being, both animal and divine. Generally, from this point of view, all created being has intrinsic ethical value, but human being may take on either special value or special obligation depending upon the theocentric tradition in question.

Nature alive is biocentric or ecocentric. It holds that value in the world resides in the natural and biotic context of which human individuals and societies are a part. Therefore, there is a natural standard of ethical rights and duties, and the good for which ethical agency and action strive can be understood in terms of systems of interdependency, relationship, sustainability, and resilience.

Obviously, I have only scratched the surface of these three orientations. The important thing to note, even with this simple sketch, is that biotechnology is interrogated by this ontological frame from the point of view of its effects on other forms of being but also from the point of view of boundaries, demarcations, and transgressions. Does biotechnology as human agency bespeak arrogance or humility? Does it represent a kind of ontological narcissism or an openness to the recognition of the being of the other? Does this new form of agency threaten to break open our traditional ontological categories altogether, radically increasing the reach of our power but also shattering the vessels of meaning within which we have traditionally understood the nature of the responsibility that accompanies power? Something like this question, I believe, was posed rather presciently about biotechnology as early as 1958 by Hannah Arendt, who wrote:

> The human artifice of the world separates human existence from all mere animal environment, but life itself is outside this artificial world, and through life man remains related to all other living organisms. For some time now, a great many scientific endeavors have been directed toward making life also "artificial," toward cutting the last tie through which even man belongs among the children of nature. . . . The question is only whether we wish to use our new scientific and technical knowledge in this direction, and this question cannot be decided by scientific means; it is a political question of the first order and therefore can hardly be left to the decision of professional scientists or professional politicians. (Arendt 1958, 2–3).

Arendt argues that it is within this web of life—as one among many "children of nature"—that human experience of the world has developed, and that we have come to comprehend not only the natural environment but also ourselves. To understand nature and life on the one hand, and meaning and humanity on the other, are not conflicting projects but are inextricably bound together. Yet the human power to create the unnatural has reached a truly radical stage, not only because it may bring about the loss of life, biodiversity, whole species on a massive scale, but also because it will fundamentally transform the human way of being in the world. How is that conceivable? It is because "humanity" is not an essence, for Arendt, but a particular condition of body and thought rooted in our connection to the earth, in what she terms our "natality."

Arendt associates "science" with this radical, and radically dangerous, transformative power. Science unlocks the inner workings of nature so that nature can be "forced," in Sir Francis Bacon's memorable metaphor, to serve human needs and desires. But again it is the step beyond Baconian science that most concerns Arendt, the step that will not merely use or manipulate nature, but replace it, particularly in the biological sphere (Lee 2005). Science as power has no value dimension; it is amoral. As such it must be directed and governed by a value-laden vision of the human good and by internal professional self-restraint and external political regulation.

What Biotechnology Teaches

I turn now from the question of how various forms of cultural meaning shape (the regulation of) biotechnology and synthetic biology to the question of how biotechnology may shape cultural and ethical meaning. Here the starting point for consideration is drawn from sources far removed from the liberal humanism or the religiosity familiar in American political culture, which tend to see the state as a contract of mutual self-interest and cooperation among free and equal natural persons. The theoretical perspective focusing on the idea of biopolitics views the state as a structure of power designed to protect the life and survival of natural bodies. The bodies (labor) and minds (creativity) of its subjects are the raw material and natural resource of the power of the state, and the primary function of state power is to manipulate and exploit (while safeguarding and investing) that resource. In the modern period, this function of making life within a political community "immunized" from perils that threaten it from outside became deliberate, strategic, and increasingly efficient, rational, and scientific (Foucault 2008, 2009; Rose 2006; Agamben 1998; Esposito 2008). The perspective of biopolitics has been described in the following way:

> That politics has always in some way been preoccupied with defending life doesn't detract from the fact that beginning from a certain moment that coincides exactly with the origins of modernity, such a self-defensive requirement was identified not only and simply as a given, but as both a problem and a strategic option. By this it is understood that all civilizations past and present faced (and in some way solved) the needs of their own immunization, but that it is only in the modern ones that immunization constitutes its most intimate essence. One might come to affirm that it wasn't modernity that raised the question of the self-preservation of life, but that self-preservation is itself raised in modernity's being [*essere*], which is to say it invents modernity as a historical and categorical apparatus able to cope with it. (Esposito 2008)

The early twentieth century anti-liberal political theorist Carl Schmidt articulated the implications of this view when he argued that the state is constituted by what he called the sovereign exception—that is, the power to exclude "mere life" from the political community for the benefit of life that served some purpose or had some communally beneficial function (Schmidt 1985).[2] For the theorists of biopolitics, this logic finds its ultimate expression in the Holocaust, Nazi eugenics, and the notion of life unworthy of life (*lebensunwertes Leben*). The flip side of this, from the perspective of those sensitive to the social control of biopolitics, is the seemingly benign and unexceptionable notion of life better than life.

Biotechnology emerges as an edifice of state/corporate power that is prone to further this state formation of control over the conditions of life and living bodies. (This is not necessarily the deliberate intention of scientists; quite the contrary.) From this point of view, philosophical accounts of value and ontology are beside the point: there are no demarcations to be made between human and nonhuman life, as both will be caught up in this discipline, this cultural formation of manipulation and control. The liberal humanistic tradition extolling individual dignity, freedom, and rights is a weak bulwark also, because biopower has already insinuated itself into those categories, colonizing the life world informed by them.

I know of no richer exploration of this facet of biopower and its effect on the very possibility of meaning than Kazuo Ishiguro's stunning novel *Never Let Me Go*, which depicts an imagined future regime of extensive organ procurement and organ transplantation, rationalized and made effective through the biotechnology of human cloning.[3] The novel follows children from the class (actually caste) of human clones as they grow up and gradually understand that they must fulfill the function for which they were created, namely, to serve as organ "donors" and "carers." Carers are clones who, before they become donors themselves, are assigned to care for those in the donor phase of their lives as they are gradually killed ("completed") by repeated organ retrievals. Ishiguro is able to explore a question that has been central to the bioethics debate concerning the (as yet speculative) future of human cloning. What would it feel like to be a clone? What would it do to one's sense of personhood, moral worth, and self-esteem to know that you were made, not begotten; created strategically for an overtly instrumental purpose?

The answers are not straightforward. The clone children in their minds rationalize the situation and generally neutralize the implications of the small bits and pieces of the truth that they are told by the "guardians" (teachers) or come to know by inference or rumor. As young adults, they struggle more as they move toward clarity and confrontation with the meaning of their situation, but they are not necessarily inclined to rebel.

Near the end of the book, a small group of clone children who went to school together and have been lifelong friends (their lives are not actually going to be that long) encounter the former head of their school. She reveals to them the scope and logic of the biopower that has come to dominate their society:

> When the great breakthroughs in science followed one after the other so rapidly, there wasn't time to take stock, to ask the sensible questions. Suddenly there were all these new possibilities laid before us, all these ways to cure so many previously incurable conditions. This is what the world noticed the most, wanted the most. And for a long time, people preferred to believe these organs appeared from nowhere...by the time they came to consider just how you were reared, whether you should have been brought into existence at all, well by then it was too late. There was no way to reverse the process. How can you ask a world that has come to regard cancer as curable, how can you ask such a world to put away that cure, to go back to the dark days? . . . So for a long time you were kept in the shadows,

and people did their best not to think about you. And if they did, they tried to convince themselves you weren't really like us. That you were less than human, so it didn't matter. (Ishiguro 2005, 262–263)

The biopower inherent in biotechnology is all the more pernicious, not because of its harms or risks of untoward results, but precisely because of the successful achievement of its promises of enhancing human security, safety, health, and function. Ishiguro reminds us of the terrible cost of dehumanization and oppression paid by all of us, and not only by those (the clones) who are oppressed. He shows that when meanings disappear from our languages of self-understanding and social construction, then our capacity to think, to act, and even to feel in ways linked to those meanings disappears likewise. Ishiguro gives us a world that stands, in his moral gaze, not so much condemned as deeply wounded and impaired. Concepts like student, guardian, giving, caring, service, possibility, completion, holding on, and letting go are all turned inside out, twisted slightly out of shape, and rendered corrupt by euphemism and double entendre. The recipients of body parts from the stockroom of the donor caste in this society gain enhanced health at the price of emaciated meaning. They may live longer thanks to their biopower, but they will not humanly prosper.

As with other modes of biotechnology and biopower, which are more problematic because they more directly manipulate the human body and spirit, some of the pioneers of synthetic biology also promise the enhancement of the human condition and the amelioration of the human estate. It is reasonable to ask what will happen if synthetic biology fails to live up to its promise. It is even more crucial to ask what could happen if it succeeds. The gift of human prospering or flourishing is elusive and multifaceted. Gains and losses are often hard to separate. When someone looks back at our age of biotechnology, will it be unambiguous to describe it using the idea of progress?

Synthetic Biology and Public Policy

I conclude with some reflections on the regulation of synthetic biology. Policy and regulation in the area of synthetic biology—as in most areas of technology and science policy and governance—will most likely focus primarily on public safety and harm issues. Secondarily, it will likely focus on discrimination and equity issues, if synthetic biology seems to put some group at risk of social harm, or if it seems to benefit one sector of society unduly and yet be developed at public expense. These issues are fundamental and important. But I want to reflect on a different set of issues, perhaps less evident, that I think are crucial and merit careful attention from policy makers.

My concern has been with the worldview associated with and exemplified in synthetic biology—or, more precisely, in the popular representation and the social reception of synthetic biology. (Of course, its popular representation and social reception may not always reflect an accurate or complete account of synthetic biology as experts see it, but it has enormous influence in its own right.) My tack has been to explore the source and shape of social concerns about biotechnology, especially from the vantage point of how it teaches us to think about natural systems, about the relationship between humans and nature, and about ourselves and our communities. Does it make us see ourselves (in Aldo Leopold's words) as "plain citizens of the biotic community"? Or as appropriators of meaningless raw biotic material? Does it teach us to see ourselves as civic trustees and stewards of a fragile and increasingly fragmented web of life? Or does it teach us to see ourselves as fabricators, improvers, exploiters, and engineers of a world that is imperfect precisely because it is very complex, fragile, and prone to eluding our deliberate control?

My general thesis is that biotechnology and synthetic biology indeed convey civic and moral narratives that have transformative effects—they have to; the notion of value neutrality in technology is a myth—and that the lessons they teach are on the whole going to be the wrong ones. When I speak of narratives here, I do not mean so much explicit stories or ideological accounts as I mean the subtle and long-acting ways in which biotechnology can tacitly alter the shape of our social institutions and our perception of our own humanity as embodied being, dependent upon a larger network of life. Such narratives have to do with formation and reformation of what the philosopher Charles Taylor has called the social imaginary, and what I have been referring to here rather simplistically as cultural meaning (Taylor 2003). This is the civic dimension of biotechnology and its moral imagination. It is not adequately described or assessed within the vocabulary of benefit, risk, safety, and harm. Nor can we measure and quantify it in the same way as we do risk and harm. The study of the relationship between innovations in biotechnology and changes in the social imaginary or our sense of what it is to be human is different from the study of, say, virology.

No doubt, some advances in synthetic biology will produce both anthropocentric and ecocentric improvements and benefits, as well as unforeseen damages and inadvertent (or intentionally evil) harms. Carefully directed synthetic applications of the capabilities of engineered organisms for use as alternative fuels, for agricultural productivity, or for bioremediation may have positive ecological and economic benefits. But when those accounts have been rendered, our policy work will not yet have been completed, and our governance responsibilities in a democracy will not yet have been met. What remains to be considered are the ways in which the civic education offered by synthetic biology (and biotechnology more broadly) undermines the task of developing an alternative to the human appropriation, manipulation, and "engineering" of nature and of life. For it is an alternative vision and set of values—this change of worldview, this cultural evolution—that we must seek if we are to take seriously our goals of sustainability, resilience, and social justice. Green engineering fixes via synthetic biology for specific environmental or economic problems, even assuming that their benefits outweigh their risks, will not get us where we need to be. How can synthetic biology be a teacher of humility? How can it provide for us the wisdom of ignorance?

Another way to put this is that synthetic biology is a technology that may temporarily solve some *problems in our lives*, but it is also a cultural formation that will contribute, almost certainly, to the further distortion of the *patterns of our lives*. It reflects a sensibility that will only further obscure a recognition that is crucial for our future: I refer to the recognition that human beings are embodied, embedded, "natured" creatures, necessarily and fundamentally. We are not the engineers of life.

Are such considerations relevant and appropriate to policy making in a pluralistic democracy? Most liberal political theorists today would say no, because we are a secular democracy. I would say yes, and all the more so because we are a secular democracy. Policy to govern science and technology that focuses exclusively on how the technology may harm us and that totally ignores considerations about how the meaning and culture of that technology may shape us is inadequate, one-eyed policy. Arguably, the civic and moral shaping of cultural meaning and worldviews are more important in a democracy than under other modes of governance. The frame used in the report of a presidential commission on bioethics in the early 1980s, *Splicing Life*, discounted and dismissed cultural concerns of the kind I am raising by saying that they really amount to a version of a religious objection to biotechnology (President's Commission 1983).[4] The religious objection cited holds that synthetic biology is morally inappropriate because it is a form of "playing God." That is to say, it is a kind of human trespass into the sacred and a kind of sinful human arrogance. This criticism, defenders of biotechnology point out, is based on special metaphysical or theological beliefs that are out of place in the discourse of public policy analysis.

But this frame of response, which defenders of biotechnology have been using against critics for many years now, is utterly inadequate for capturing the significance of the concerns I am trying to articulate here. Religious notions of the right relationship between humans and the creation certainly are one basis for concerns about biotic or ecological citizenship, trusteeship, and stewardship; but they are not the only basis, and it does not follow that these concerns are "religious" or metaphysical arguments. To think so is to overlook the ways in which these concerns are in fact grounded in the best science we have today as regards such fields as ecology, conservation biology, and geophysical fluid dynamics. To be a synthetic biologist is not to usurp the place of God (theologically a very bizarre notion, when you think about it), or even, pathetically, to try, but it may well be to manipulate life in ways that are of dubious wisdom from a social and a scientific point of view.

To this one might respond: "Synthetic biology works, doesn't it?" Is the fact that something "works" the final determinant of its scientific legitimacy? I don't think so, any more than it is the final determinant of ethical legitimacy. Synthetic biologists have hit upon a level of genomic and biological functioning (at very elemental and small levels) that they can seemingly control rather impressively. So synthetic biology does "work" in a sense and therefore represents scientific discovery of a new truth in the biological sciences. Any yet, despite this, I have a feeling that synthetic biology's manipulations are profoundly out of step with the current best thinking in biology, in that they seek to dispense with, rather than to understand, the complexity and holistic properties of biological systems at all levels (Noble 2006).

I acknowledge that these reflections do not entail a particular set regulations concerning the future conduct of synthetic biology. But if we take one step back, I believe this analysis does suggest some direction for the conduct of the policy-analysis and ethics discourse that must precede specific policies and regulations for synthetic biology. And perhaps this somewhat broader vantage point should be seen as an integral part of any policy analysis, for surely it is only one part of the task of policy analysis and bioethics to suggest concrete regulations, not their be all and end all.[5] Another part of that task, especially in instances of significant technological innovation with potentially far-reaching institutional and economic consequences, is to consider carefully the foundations of the normative legitimacy of policy governance, and the structural prerequisites of appropriate and effective governance in case there are features of the new technology that exceed the current capacity of the democratic regulatory apparatus itself. Finally, a third key aspect of policy analysis and ethics, to which I have tried to contribute, may be to warn against misplaced technological promise and subtle misdirection.

In my judgment, the policy analysis and ethics of synthetic biology thus far have not adequately addressed this range of issues. Yet, won't that shortcoming be remedied by more study and more regulatory oversight, so that if a current approach is too permissive in light of future experience, it can be adjusted and made more stringent as needed? Not necessarily. That seemingly prudent and sensible regulatory scenario rests on familiar assumptions that may not serve us well in the case of synthetic biology. If background questions of (a) the cultural formation of moral and political legitimation and (b) the structural limitations of governance are not explicitly and systematically examined in a policy analysis, then regulations may only touch the surface of the social effects of a new technology. And superficial or overly narrow regulatory strategies, while they may guard against a particular risk (such as laboratory accidents) or while they may facilitate research and development (by providing access to sources of public funding and private investment), may also inadvertently have the long-run effect of fostering and exacerbating uses of the technology, attitudes toward it, and concentrations of corporate power and control over it that are deleterious in themselves and also corrosive of the democratic governance of science and technology.

In sum, I am calling for a type of policy analysis that includes a careful background consideration of legitimation formation and the structure of governance. The difficulty and challenge of this type of policy analysis should not be underestimated. In order to achieve it, both our policy process and our broader conversation of democratic citizenship will have to change, and this is an interdisciplinary responsibility in which both scholars and activists have a role to play. Doing this now may help to prepare us for handling new domains of application for synthetic biology, such as the domain of human health or, perhaps, enhancement. It may also prepare us for dealing with novel forms of biotechnology in the future; for synthetic biology will not be its last innovation.

Where will this reconstructed discourse of normative policy analysis and the ethics of biotechnology come from and how can we develop it? I close with four suggestions along those lines.[6]

1. *Carefully conceptualize the nature of the power synthetic biology offers. Find a way of naming that power as such. Avoid the rhetoric of*

radical newness, which leads to fear but also to allure, and the opposite rhetoric of unexceptional continuity, which leads to complacency and reification. It is vital to continue to ask Francis Bacon's fundamental question about science and technology: How should synthetic biology be understood as a practice of human power? (Jonas 1974) Is it a breathtaking new extension of human ability to control and create biological processes to achieve human goals, or is it an overreaching and simplistically reductionist paradigm that will fall short of the hyperbole surrounding it? Critics of synthetic biology can easily be tempted to adopt the rhetoric of newness, strangeness, and transgression in order to arouse public concern, just as the proponents of synthetic biology are using that rhetoric to gain financial backing and public support (although they cast the transgression in heroic Promethean images, rather than darker Faustian ones). But if critics reinforce this public understanding concerning synthetic biology, then they face the dilemma (and paradox) of undermining some of their most telling normative objections to this form of biotechnology, namely that it is unnatural and out of keeping with biological constraints and evolutionary complexity.

A closely related question concerns how novel or how humdrum both critics and supporters want synthetic biology to seem. Its novelty brings excitement. But its continuity with tried-and-true techniques of recombinant DNA research, and even traditional forms of plant and animal breeding, make it seem much safer and unexceptional. Just how new is the field of synthetic biology, and how distinct is it from older genetic engineering techniques? What are the new risks, and what aspects of the technology are simple extensions of existing biomedical engineering? How does synthetic biology relate to human biological modification more generally?

2. *Develop a critical perspective on biotechnology and synthetic biology from within a progressive, egalitarian, and democratic socialist tradition.* Earlier I identified three influential critical perspectives (counternarratives) on biotechnology: the precautionary (obligations to practice prudence and humility), the liberal humanist (obligations to respect human dignity and freedom), and the ontological (obligations to respect nature). These normative frameworks are spread across a broad religious, cultural, and political spectrum, but thus far they have been articulated more strongly by voices that identify themselves with religious and cultural conservatism than by voices on the secular left. During the nineteenth and twentieth centuries, political theories and social movements of the left mainly took a positive stand on the advancement of science and

technology because they believed it could be used to equalize wealth and power in society and to improve the quality of life of classes of people who had traditionally been deprived and compelled into alienating and unhealthy labor to earn a living (Macpherson 1973). The poor ecological record of state socialism prior to 1989 and the advent of new kinds of environmental problems, together with biotechnological advances themselves, make that socialist framework inadequate and obsolete today. Can a progressive, left critique of biotechnology emerge on a different footing?

I believe there are reasons to be optimistic on this score. Progressive intellectuals and scholars influence society and politics by reframing central concepts and categories of ideological conflict and lived experience. Major shifts have taken place in this way during the past thirty years on the concepts of race, gender, and self-identity (Rogers 2011). The time may be right for an analogous sea change in the orientation that intellectuals take toward biotechnology. Virtually all discussions of biotechnology—and most pointedly, synthetic biology—are framed by conventional conceptual logics regarding science, nature, control, health, risk, and human well-being. Limited political progress can be made within the confines of these conceptual logics. If a more egalitarian, democratic, and social and environmental justice–oriented biopolitics is to emerge globally and within the advanced capitalist nation-states, then conceptual reframing work must be enabled and supported in the intellectual domain.

Such turnings in the past have been made possible through mentoring and fellowship support for young scholars, conferences aimed specifically at exploring interdisciplinary discourse and philosophical critique, the orientation of intellectual journals toward this task, the creation of book series by university presses or small independent presses to support and disseminate these ideas, and now a host of web-based modes of publication and intellectual exchange. Initially difficult, arcane, and accessible primarily to a small audience of sophisticated and highly motivated participants, such an intellectual orientation can gradually gain momentum and can filter out into much broader domains over a period of time. It can form a touchstone and a reference point for more direct modes of policy engagement, movement building, and protest actions.

3. *Bring the discourses of ethics in the realm of genetics, human health, and molecular science into closer dialogue with the discourse of ethics and science in the areas of ecology, conservation biology, and environmental policy.* At this time, remarkably little dialogue takes place between the molecular side of biology and the ecosystemic and evolutionary sides of

the life sciences. What is the relationship between biotechnology and factors such as climate change, biodiversity loss, and the biophysical limits of the planet; how can these technologies be employed in the development of more renewal energy sources and sustainable economic systems?

4. *Place the regulation of the science and technology of biotechnology in the broader context of the regulation of the political economy of biotechnology, and see this not merely as a problem for governance within each nation-state but as a problem of global governance.* Biotechnology poses an ethical and a regulatory challenge not merely because it is extending the capacities of human manipulation and control to the molecular level of life, but also because it is forming around itself a global economic structure of corporate control, speculative investment, and transnational communication and sharing of expertise. The do-it-yourself or "garage science" aspects of synthetic biology have been widely noted, and the physical infrastructure necessary for the use of this technology is of course one important facet of its regulatory governance and public control. But no less of a challenge for governance looms in the opposite direction, namely, the concentrated financial and corporate resources standing behind synthetic biology promoting its development and its global reach. Vested interests in promoting synthetic biology will comprise a formidable lobbying force in domestic politics, making strong public regulation problematic. And inadequate or undeveloped international legal and regulatory mechanisms make transnational traffic in synthetic biology very difficult to control.

As if these factors did not make democratic governance of synthetic biology difficult enough, an additional factor is the growing collaboration between the corporate bioeconomy's interest in synthetic biology and that of the national security bureaucracy, particularly in the United States. The extraordinary legal secrecy and insulation provided the national security apparatus from ordinary democratic oversight and governance is something that future policy analysis and bioethics will have to take into consideration. Whether effective regulation of synthetic biology in the context of national security interests is possible remains to be seen. It will surely require a large groundswell of civil society and grassroots concern and demand if it is to come about.

Often in the past, bioethics has been slow to take up the economic context of science, medicine, and technology. When the economic factor has been addressed from an ethical perspective, more attention has been given to individual misconduct and conflict of interest than to structural features of power, direction, and governance that affect the public interest. Changing this focus is one of the strong suits of the kind of critical policy studies, social science, and bioethics that builds on the conceptual frameworks and intellectual traditions of the progressive left. That disposition and contribution will be essential to the proper development of a policy ethics for synthetic biology.

These, then, are four of the several kinds of new conceptual and political work that should be done as bioethics confronts biopolitics and bioeconomics in the guise of synthetic biology. We must have new viewpoints emerging from the humanities and the social sciences, and from fiction, film, and the other arts, so that these connections can be finely understood and richly imagined. Without these resources of value, narrative, and meaning, we will not be able as democratic citizens to meet the governance challenge that the brave new world of biotechnology poses for us. Grasping promise and avoiding peril are the tasks of democratic prudence; retaining a sense of our proper place in the world of life and fulfilling our responsibilities for that world are the tasks of democratic principle; and they are its calling.

Notes

1. Weber was concerned with linking the rise of capitalism with cultural and psychological reactions to the doctrine of predestination and the desire to influence conditions of personal salvation; my focus here is on a different aspect of this cultural tradition.

2. See also Agamben (1998, 15–29). For a discussion of how liberalism might respond to the problem of biopower, see Jennings (2003).

3. These next few paragraphs draw upon my previously published essay Jennings (2010).

4. A similar framework of argument informs the recent work of the Presidential Commission for the Study of Bioethical Issues (2010) A contrasting perspective can be found in the European Group on Ethics in Science and New Technologies to the European Commission (2010).

5. I have discussed this perspective on policy analysis in a number of previous essays (Jennings 1982, 1987, 1988).

6. For the development of the following ideas, I am particularly grateful for extended conversations with Richard Hayes, Marcy Darnovsky, David Winickoff, Mark Brown, Mary Shanley, Osagie Obasogie, Daniel Sharp, and Emily Smith Beitiks.

References

Agamben, Giorgio. 1998. *Homo Sacer: Sovereign Power and Bare Life*. Stanford: Stanford University Press.

Arendt, Hannah. 1958. *The Human Condition*. Chicago: University of Chicago Press.

Bauman, Zygmunt. 2000. *Liquid Modernity*. Cambridge, UK: Polity Press.

Bell, Daniel. 1976. *The Cultural Contradictions of Capitalism*. New York: Basic Books.

Collins, H. M., and Robert Evans. 2002. The third wave of science studies: Studies of expertise and experience. *Social Studies of Science* 32 (2):235–296.

Dworkin, Ronald. 2000. Playing God: Genes, clones, and luck. In *Sovereign Virtue: The Theory and Practice of Equality*, ed. Ronald Dworkin. Cambridge, MA: Harvard University Press.

Esposito, Roberto. 2008. *Bios: Biopolitics and Philosophy*. Minneapolis: University of Minnesota Press.

European Group on Ethics in Science and New Technologies to the European Commission. 2010. *Ethics of Synthetic Biology*. Luxembourg: Publications Office of the European Union.

Evans, J. H. 2002. *Playing God? Human Genetic Engineering and the Rationalization of Public Bioethics Debate*. Chicago: University of Chicago Press.

Foucault, Michel. 2009. *History of Madness*. New York: Routledge.

Foucault, Michel. 2008. *The Birth of Biopolitics: Lectures at the Collège de France 1978–1979*. New York: Palgrave Macmillan.

Fukuyama, Francis. 1995. *Trust: The Social Virtues and the Creation of Prosperity*. New York: Free Press.

Fukuyama, Francis. 2002. *Our Posthuman Future: Consequences of the Biotechnology Revolution*. New York: Farrar, Straus, and Giroux.

Giddens, Anthony. 1976. *New Rules of Sociological Method*. New York: Basic Books.

Habermas, Jürgen. 2003. *The Future of Human Nature*. Cambridge, UK: Polity Press.

Ishiguro, Kazuo. 2005. *Never Let Me Go*. New York: Vintage Books.

Jennings, Bruce. 1982. Interpretive social science and policy analysis. In *Ethics, The Social Sciences and Policy Analysis*, ed. Daniel Callahan and Bruce Jennings, 3–35. New York: Plenum Press.

Jennings, Bruce. 1987. Interpretation and the practice of policy analysis. In *Confronting Values in Policy Analysis: The Politics of Criteria*, ed. Frank Fischer and John Forester, 128–152. Newbury Park, CA: Sage Publications.

Jennings, Bruce. 1988. Political theory and policy analysis: Bridging the gap. In *Handbook of Political Theory and Policy Science*, ed. Edward B. Portis and Michael B. Levy, 17–28. Westport, CT: Greenwood Press.

Jennings, Bruce. 2003. The Liberalism of Life: Bioethics in the Face of Biopower. *Raritan Review* 22 (4):130–144.

Jennings, Bruce. 2010. Biopower and the romance of liberation. *Hastings Center Report* Jul-Aug:1–4.

Jonas, Hans. 1974. Seventeenth century and after: The meaning of the scientific and technological revolution. In *Philosophical Essays: From Ancient Creed to Technological Man.*, 45–80. Englewood Cliffs, N.J.: Prentice Hall.

Kass, Leon R. 2002. *Life, Liberty and the Defense of Dignity: The Challenge for Bioethics*. San Francisco: Encounter Books.

Kass, Leon R., and James Q. Wilson. 1998. *The Ethics of Human Cloning*. Washington, DC: AEI Press.

Kennedy, John F. 1962. Remarks to members of the White House Conference on National Economic Issues, May 21, 1962. http://www.jfklink.com/speeches/jfk/publicpapers/1962/jfk203_62.html.

Lasch, Christopher. 1978. *The Culture of Narcissism*. New York: Norton.

Lee, Keekok. 2005. *Philosophy and Revolutions in Genetics: Deep Science and Deep Technology*. London: Palgrave Macmillan.

Lewis, C. S. [1947] 1955. *The Abolition of Man*. New York: Macmillan.

Macpherson, C. B. 1973. *Democratic Theory: Essays in Retrieval*. Oxford: Oxford University Press.

Mill, John Stuart. 1985. Nature. In *The Collected Works of John Stuart Mill*. Vol. 10, *Essays on Ethics, Religion, and Society*, ed. John M. Robson, introduction by F. E. L. Priestley. Toronto: University of Toronto Press; London: Routledge and Kegan Paul, 373–402.

Mitchell, C. G., E. D. Pellegrino, J. B. Elshtain, F. F. Kilner, and S. B. Rae. 2007. *Biotechnology and the Human Good*. Washington, DC: Georgetown University Press.

Noble, Denis. 2006. *The Music of Life: Biology beyond the Genome*. New York: Oxford University Press.

Nussbaum, Martha C., and Cass Sunstein, eds. 1998. *Clones and Clones: Facts and Fantasies about Human Cloning*. New York: Norton.

Presidential Commission for the Study of Bioethical Issues. 2010. *New Directions: The Ethics of Synthetic Biology and Emerging Technologies*. Washington, DC: Presidential Commission for the Study of Bioethical Issues.

President's Commission for the Study of Ethical Problems in Medicine and in Biomedical and Behavioral Research. 1983. *Splicing Life*. Washington, DC: U.S. Government Printing Office.

President's Council on Bioethics. 2002. *Human Cloning and Human Dignity*. New York: Public Affairs.

Raffensperger, Carolyn, and Joel Tickner (eds.). 1999. *Protecting Public Health and the Environment: Implementing The Precautionary Principle*. Washington, DC: Island Press.

Ricoeur, Paul. 1970. *Freud and Philosophy: An Essay on Interpretation*. Trans. Denis Savage. New Haven: Yale University Press.

Rogers, D. 2011. *Age of Fracture*. Cambridge, MA: Harvard University Press.

Rose, Nicholas. 2006. *The Politics of Life Itself: Biomedicine, Power, and Subjectivity in the Twenty-first Century*. Princeton: Princeton University Press.

Sahlins, Marshall. 2010. Infrastructuralism. *Critical Inquiry* 36 (3):371–385.

Sandel, Michael. 2007. *The Case against Perfection: Ethics in the Age of Genetic Engineering*. Cambridge, MA: Harvard University Press.

Schmidt, Carl. 1985. *Political Theology: Four Chapters on the Concept of Sovereignty*. Cambridge, MA: MIT Press.

Taylor, Charles. 2003. *Modern Social Imaginaries*. Durham, NC: Duke University Press.

Weber, Max. (1905) 1958. *The Protestant Ethic and the Spirit of Capitalism*. New York: Charles Scribner's Sons.

9

"Teaching Humanness" Claims in Synthetic Biology and Public Policy Bioethics

John H. Evans

The emergence of synthetic biology has led some commentators to raise some deep and abstract ethical issues. It has been repeatedly claimed, for example, that synthetic biology will teach people to accept a different notion of what it means to be human—a view I will call the teaching humanness (TH) claim. TH claims have been made about various scientific innovations over the centuries, as well as in more recent debates in public policy bioethics, where such claims are routinely made and then ignored. In this paper I first discuss why claims like this are ignored in public policy bioethics, with one reason being that people are unsure how to evaluate their legitimacy. I then describe the claim for synthetic biology and show that TH claims are pervasive in public policy bioethics. I then show that these claims are of critical importance, and that we should act on TH claims, if they are legitimate. I then describe a staged screening method for evaluating TH claims, and using the method I conduct a preliminary analysis of the claims about synthetic biology. I find little support for one TH claim and am unable to reach a clear conclusion on the other. I conclude with a discussion of what our ethical options would be were a TH claim be found to be legitimate.

A Typology of Claims in Public Policy Bioethics

Public policy bioethics is the discourse of professionals, primarily academics, about what society should do about a scientific innovation. The goal of this mediating institution in the public sphere is to influence the ethical content of public policy (Evans 2012). A classic public policy bioethics question is: should the NIH pay for germline human genetic engineering? This type of bioethics is distinct from research bioethics and health care ethics consultation, which concern the ethics of scientific research and health care settings, respectively. The ethics of synthetic biology is a typical public policy bioethics topic.

Synthetic biology attempts to create new forms of life purely from design, instead of combining components of pre-existing life-forms (van Est, de Vriend, and Walhout 2007). In one of its ideal-type technologies, the aim is to create a "minimal bacterial genome" that has the absolute minimal number of genes required for continued existence and reproduction. On this "chassis" would be attached various artificially produced genomic "cassettes" that would create an organism with desired properties, such as producing fuel or the components of medicine. Unlike the previous technology of genetic engineering, where genes from a fish are inserted into the genes for a tomato, the newly invented life-form would be almost purely human artifice. It has been claimed that this genetic manipulation is so much more powerful than previous methods that it appears to produce qualitatively different ethical problems (Boldt and Müller 2008).

The Emerging "Thin" Debate over Synthetic Biology

In the emerging ethical debate about synthetic biology, the vast majority of the stated ethical concerns have been what I have elsewhere called "thin." "Thin" claims in public policy bioethics have many features (Evans 2002, 14–21), but for now I will focus on the fact that "thin" claims do not produce a debate about which values we as a society should be pursuing, but rather they assume universally held values and then talk of how to avoid violating those values. To avoid any debate about values, thin claims typically base their claims, explicitly or implicitly, on what they purport to be the universal morals of Western civilization (Evans 2012), usually referred to as "the common morality" (Turner 2003; Beauchamp 2003).

The emerging ethical claims about synthetic biology are thin and focused on defending such values. For example, some claim that synthetic biology could potentially kill people as synthesized life-forms get out of control, violating the value of nonmaleficence. An ethical study by an institute that engages in synthetic biology research "focuses on three key societal issues: bioterrorism . . . worker safety . . . and protection of communities and the environment in the vicinity of legitimate research laboratories" (Garfinkel et al. 2007, 8). Similarly, a European research institute report focuses upon "biosafety," "misuse and bioterrorism," and "intellectual property" (van Est, de Vriend, and Walhout 2007). Intellectual property is probably considered a "justice" issue, but the rest are concerned with nonmaleficence. The first published ethical statement on synthetic biology of which I am aware included concerns that synthetic biology could "harm our health or the environment," "intellectual property and commercialization issues," and fear of "the creation or release of organisms that could be used as biological weapons" (Cho et al. 1999). Boldt and Müller summarize this "thin" debate well:

> Ethical considerations are concerned, first and foremost, with potentially harmful, unintended effects of new technological abilities and, where applicable, with possible intentional abuse of the technology in question. Accordingly, statements and discussion to date referring to the ethical dimensions of synthetic biology have focused on risk evaluation and management. In the case of synthetic biology, specific risks in need of close scrutiny and monitoring are uncontrolled self-replication and spreading of organisms outside the lab, and deliberate misuse by terrorist groups or individuals or by "biodesigner-hackers." (2008, 387)

Of course these are legitimate ethical questions that should be debated. However, if the history of public policy bioethics debate is a guide, there are additional claims, which I will call "thick," that will be raised by scholars who are typically at the margins of the influential debate, and these "thick" claims will either be dismissed through various mechanisms, like labeling them "religious," or by translating them into thin claims while not acknowledging what is lost in the translation. "Thick" concerns are not about defending the assumed and undiscussable values of the "thin" debates, but rather concern debating what our society's values should actually be. For example, in earlier debates about human genetic engineering, there was a "thick" debate about the values we would use to decide what sort of human to create (Evans 2002). Thick debate for many reasons appears to critics to be vague, and is often dismissed on those grounds. Typical is an early dismissal of these sorts of claims in synthetic biology debates, wherein Bedau and Triant reject the "intrinsic objections" that synthetic biology is a violation of nature, leads to the commoditization of life, or is "playing God." They conclude that since "all of the intrinsic objections to the creation of artificial cells canvassed in this section turn out to be vague, simplistic, or ill-conceived," people should turn to the serious concerns, which are based on a utilitarian maximization of human happiness—which is one of the universal values presumed by common morality theories (Bedau and Triant 2009, 35).

The early dominance of "thin" claims in the synthetic biology debate can be explained by a number of factors. First is the involvement of mainstream bioethicists, who, as I have discussed elsewhere at length, avoid the "thick" concerns (Evans 2012). A second reason is the early involvement of scientists in trying to frame this ethical debate as it emerges. The first ethical statement about synthetic biology was funded by the institute of one of the main scientists in this enterprise (Cho et al. 1999, 2090), and this same institute has produced other ethical and policy analyses of the issue (Garfinkel et al. 2007). While individual scientists obviously have a broader ethical repertoire, scientific institutions tend to recognize only a very limited number of values to forward—essentially the relief or avoidance of human suffering and the discovery of knowledge. They then view these values, which happen to coincide with the skill set that scientists bring to the debate, as the primary values to consider.

Finally, a third reason "thick" ethical claims are ignored or translated is that they are often made in a loose language akin to what Gustafson has called "prophetic discourse," as compared to the more precise "ethical" or "policy" discourse (Gustafson 1990). Although our elected officials act all the time on arguments that are made at the level of prophetic discourse, it is probably easier for them to act when claims are made with a greater degree of specificity. It certainly is for unelected government officials. So, for these important "thick" claims to be taken seriously, a method to specify them more precisely must be developed. This paper endeavors to do exactly that.

"Thick" Claims about Synthetic Biology

Many, but not all, of the ethical reports on synthetic biology gesture toward the thick concerns. For example, the European research institute report gestures toward the thick in one paragraph titled "What is life?" The paragraph is simply a list of questions:

> May we intervene in life? What is our definition of life? To what extent do we still view artificial biological systems as "living material"? How does a completely artificial cell differ from a machine? What criteria do we use to determine to designate life? Do we restrict the principle of "worthy of protection"; to natural organisms, to (evolved) creation, or do we consider artificial forms of life also worthy of protection? (Van Est, de Vriend, and Walhout 2007, 13)

Answering these questions would require a determination of values, as in, What is life? or, What do we want "life" to be? If history is our guide, these questions will eventually be ignored in favor of addressing the "thin" concerns.

A Recurring "Thick" Claim

Many of the early participants in the synthetic biology debate have raised the same "thick" claim—that synthetic biology may teach the public an undesirable conception of what it means to be human. This is the teaching humanness (TH) claim, and I will show that this claim is made repeatedly in bioethical debates. By *teaching*, I mean that humans have a particular mental construct, and are taught a new one that replaces the old. By *humanness*, I mean theories of what it means to be human. For example, at its most broad, a Christian notion of humanness holds that we are human because we are made in the image of God. A biological notion holds that we are human because we have the human DNA sequence. This is the ideal-type "thick" concern because "what we humans should think ourselves to be" is the most value-laden question we can imagine, and therefore serves as a vehicle to develop our thinking about how public policy bioethics can handle "thick" concerns. I will discuss two versions of the TH claim made about synthetic biology.

The first version of the claim is made in the first ethical statement on synthetic biology, where Cho and her colleagues articulate the fear that the current use of the term *life* to discuss both bacteria and humans teaches the public that "life" for bacteria is nothing more than DNA, and so people will make the category mistake and conclude that human "life" is nothing more than DNA. In other words, people will be taught by this technology that humans are just chemicals, and can be treated the way we treat bottles of chemicals. They write, "What are the ultimate implications of defining life in terms of DNA? . . . There is a serious danger that the identification and synthesis of minimal genomes will be presented by scientists, depicted in the press, or perceived by the public as proving that life is reducible to or nothing more than DNA. . . . This may threaten the view that life is special" (Cho et al. 1999, 2089).

Boldt and Müller make the same claim. They start with the point that synthetic biology is the first technology to truly give humans the ability to create new life-forms instead of just manipulating existing ones. They write that "from an ethical perspective, this shift from 'manipulatio' to 'creatio ex existendo' is decisive because it involves a fundamental change in our way of approaching nature" (Boldt and Müller 2008, 388). The problem with this is pedagogical—it teaches us something about us humans. They continue: "As a consequence, the way in which newly created organisms are conceptualized has an ethical impact on how life in general is understood and valued. When describing microorganisms and their signaling pathways, synthetic biology researchers often invoke the computer metaphor of 'hardware' and 'software.'" This leads to the underlying problem: "All of this vocabulary identifies organisms with artifacts, an identification that, given the connection between 'life' and 'value,' may in the (very) long run lead to a weakening of society's respect for higher forms of life that are usually regarded as worthy of protection" (Boldt and

Müller 2008, 388). Again, as with Cho and her colleagues, the concern is that synthetic biology will lead to people learning a particular definition of bacterial "life," which they then extend by analogy to human "life." Humans will be taught to think of each other as more objectlike. The harm is, implicitly, we will then treat each other like we treat objects. I will call this the "teaching a biological humanness" version of the TH claim.

Boldt and Müller paradoxically make a second TH claim that is the radical inverse of thinking of ourselves as only DNA. That is, we will come to think of ourselves as gods. They write that we will "come to understand ourselves, justified or not, as creators of 'animuncula' (a term that we think captures very nicely the associations that accompany the creation of new forms of life), with all the implications that such a self-understanding might bring with it" (Boldt and Müller 2008, 388). "The human self-conception" will be "as 'creator'" (Boldt and Müller 2008, 388).

Similarly, Bedau and Parke write that synthetic biology may "resolve one of the remaining fundamental mysteries about creation and our place in the universe." It may have unpredictable implications for "our understanding of ourselves and the world we live in." Therefore, some people may advocate "that society should be consulted before it is burdened with the need to refashion its sense of itself" (Bedau and Parke 2009, 651). Let us call this the "teaching a divine humanness" version of the TH claim. In this essay I take these TH claims on their face, that a transformation in our sense of humanness is undesirable and should be opposed.

Teaching Humanness in Public Policy Bioethics

Erik Parens and his colleagues have recently pointed out that synthetic biology raises no issues that have not been addressed by previous bioethical discussions. I agree, and also agree with their point that just because an issue has been discussed in the past does not mean that its new presentation isn't a good opportunity to raise it again and see if we can get a better understanding of the issue (Parens, Johnston, and Moses 2008).

The argument that humans used to have a certain self-perception, but were taught a different one by scientists, has been repeatedly made not only in the history of public policy bioethics, but in the history of public interaction with science. For example, it is classically said that the Copernican revolution changed human self-conception, that previous to the Copernican vision,

> in medieval thought, man occupied a more significant and determinative place in the universe than the realm of physical nature, while for the main current of

modern thought, nature holds a more independent, and more permanent place than man. . . . [In] the Middle Ages man was in every sense the centre of the universe. The whole world of nature was believed to be teleologically subordinate to him and his eternal destiny. (Burtt 1959, 4)

Moving ahead a few hundred years, the same accusation has been repeatedly been made about Darwinism. For example, in the 1920s William Jennings Bryan, of Scopes "monkey trial" fame was opposed to teaching evolution because it violated his literalist Genesis narrative. He was also concerned that Darwinism teaches humans, at minimum, a form of morality, and at maximum, a notion of humanity. According to Ronald Numbers's canonical analysis of creationism, for Bryan:

World War I . . . . exposed the darkest side of human nature and shattered his illusions about the future of Christian society. Obviously something had gone awry, and Bryan soon traced the source of the trouble to the paralyzing influence of Darwinism on the conscience. By substituting the law of the jungle for the teachings of Christ, it threatened the principles he valued most: democracy and Christianity. Two books in particular confirmed his suspicion. The first . . . recounted first hand conversations with German officers that revealed the role of Darwin's biology in the German decision to declare war. The second . . . purported to demonstrate the historical and philosophical links between Darwinism and German militarism. (Numbers 1982, 538)

This theme continues today with intelligent design advocates, who say that the materialist conception of reality promoted by Darwinism

eventually infected virtually every area of our culture, from politics and economics to literature and art. . . . [M]aterialists denied the existence of objective moral standards, claiming that environment dictates our behavior and beliefs. Such moral relativism was uncritically adopted by much of the social sciences.... Materialists also undermined personal responsibility by asserting that human thoughts and behaviors are dictated by our biology and environment. . . . In the materialist scheme of things, everyone is a victim and no one can be held accountable for his or her actions. (Discovery Institute, n.d.)

Social scientists have also claimed that modern biology has taught the public the idea of genetic essentialism that we "are" our genes (Nelkin and Lindee 1995; Condit, Ofulue, and Sheedy 1998; Condit 1999). As one article summarizes the debate: "For several decades, scholars have expressed concern that the growth in our knowledge and understanding of human biology pushes us even further into a reductionistic worldview in which human agency, social structure, culture, and free will are erased by deterministic formulas that describe human beings as mere animals respecting the iron laws of physics and evolution" (Condit, Ofulue, and Sheedy 1998, 979).

Similarly, the sociologist Howard Kaye (1997) writes that "the aim of current [scientific] efforts is . . . to transform the human self-conception by translating our lives and history back into the language of nature so that we might once again find a cosmic guide for the problems of living" (p. 3). When thinking of the body, this view encourages us to think of the body as nothing more special than a series of interconnected biological devices. Kaye summarizes the underlying notion of humanness behind the recent genetic revolution:

> In both aim and impact, the end of this revolution is a fundamental transformation in how we conceive of ourselves as human beings and how we understand the nature and purpose of human life rightly lived. . . . we are in the process of redefining ourselves as biological, rather than cultural and moral beings. Bombarded with white-coated claims that "Genes-R-Us," grateful for the absolution which such claims offer for our shortcomings and sins, and attracted to the promise of using efficient, technological means to fulfill our aspirations, rather than notoriously unreliable moral or political ones, the idea that we are essentially self-replicating machines, built by the evolutionary process, designed for survival and reproduction, and run by our genes continues to gain. (Kaye 1998, 488)

The feminist sociologist Barbara Katz Rothman argues that new reproductive technologies will teach us that humans—or at least the female half of the species—are objects, in this case, containers, and that people have differential value:

> The "preciousness" of the very wanted, very expensive baby will far outweigh the value given to the "cheap labor" of the surrogate.
>
> We are encouraging the development of "production standards" in pregnancy—standards that will begin with the hired pregnancy, but grow to include all pregnancies. This is the inevitable result of thinking of pregnancy not as a relationship between a woman and her fetus, but as a service she provides for others, and of thinking of the woman herself not as a person, but as the container for another, often more valued, person. (Rothman 1989, 244)

We can also see this claim in debates more central to public policy bioethics. On the day I started writing this essay, I got a fund-raising e-mail from Bioedge (5/12/09), an Australian compiler of bioethics-oriented media stories, that said, "There's very little doubt that technology is changing the way that we human beings look at each other—and it's not always for the best." In public policy bioethics the TH claim is often summarized by the term "dehumanization." The claim of dehumanization implies simply that the author has a preferred notion of the human, and believes that a scientific technology will teach people to reject that notion ("de-humanization") and subsequently adopt the objectionable notion.

For example, Leon Kass, while discussing biotechnology, writes that "unlike a man reduced by disease or slavery, the people dehumanized à la *Brave New World* are not miserable, don't know that they are dehumanized, and, what is worse, would not care if they knew" (Kass 1985, 35). The reference to the novel *Brave New World* points to a society whose entire notion of the human has been changed, and whose members therefore have no way of knowing that they are less than human, by Kass's account. This is a difficult concept, so I will offer another quote from Kass. The greatest problem with new biological technologies is our "willing dehumanization—dehumanization not directly chosen, to be sure, but dehumanization nonetheless—as the unintended yet often inescapable consequence of relentlessly and successfully pursuing our humanitarian goals" (Kass 1985, 31). Kass makes a very similar claim about organ selling:

> We wonder whether freedom of contract regarding the body, leading to its being bought and sold, will continue to make corrosive inroads upon the kind of people we want to be and need to be if the uses of our freedom are not to lead to our willing dehumanization. . . . Selling our bodies, we come perilously close to selling out our souls. There is a danger in contemplating such a prospect—for if we come to think about ourselves like pork bellies, pork bellies we will become. (Kass 1992, 82–83)

Note again that it is not that those who sell their organs are dehumanized, although Kass would probably think that too. Rather, what he sees as more important is that "even contemplating such a prospect," even debating it in our public discourse, changes our perception of humanity. In Kass's terms, we could consider those who sell their organs to be equally human as everyone else, with everyone having a debased humanity more like "pork bellies" than what we had before we allowed the sale of organs.

Also consider the theologian Gilbert Meilaender, testifying before the National Bioethics Advisory Commission on human cloning. He claimed that because of the instrumental motivation of the parents to clone, the cloned child becomes their "project" and "when the sexual act becomes only a personal project, so does the child." The instrumentality involved in our personal projects makes procreation more like reproduction. "Human cloning would be a new and decisive turn on the road" away from our previous view of children toward "an understanding of the child as a product of human will" instead of a product of the mystery of the random chances associated with old-fashioned procreation (Meilaender 1997, 42–43). We come to think of children—of humans—as more objectlike than before.

On the other end of the ideological spectrum in public policy bioethics from Kass and Meilaender, Arthur Caplan similarly claims that markets in organs "convert human beings into products, a metaphysical transformation that cheapens the respect for life and corrodes our ability to maintain the stance that human beings are special, unique and valuable for their own sake, not for what others can mine, extract, or manufacture from them" (Caplan 1997, 100). That is, the mere knowledge of markets in organs will change our notion of the human—the "metaphysical transformation." Although Caplan does not explicitly state what he thinks a human is, or should be, he clearly believes that the transformation caused by selling body parts is not good. It "cheapens the respect for life" and teaches us that we are not "special, unique and valuable for our own sake."

In the examples above, TH is portrayed as the unintended side effect of scientists' activity. Others see it as the intended goal of scientific activity. For example, Joachim Schummer claims that since the reasons why we should engage in synthetic biology are so weak, "the leading motivation is to prove the 'creative power of man,' a symbolic act in the imagined rivalry with a metaphysical agency" (Schummer 2009, 137). Similarly, Paul Ramsey wrote of human genetic engineering in 1970 that scientists were trying to displace God, and thus teach us the meaning of life:

> Taken as a whole, the proposals of the revolutionary biologists, the anatomy of their basic thought-forms, the ultimate context for acting on these proposals provides a propitious place for learning the meaning of "playing God"—in contrast to being men on earth.
>
> [The scientists have] "a distinctive attitude toward the world," "a program for utterly transforming it," an "unshakable," nay even a "fanatical," confidence in a "worldview," a "faith" no less than a "program" for the reconstruction of mankind. These expressions rather exactly describe a religious cult, if there ever was one—a cult of men-gods, however otherwise humble. These are not the findings, or the projections, of an exact science as such, but a religious view of where and how ultimate human significance is to be found. It is a proposal concerning mankind's final hope. One is reminded of the words of Martin Luther to the effect that we have either God or an idol and "whatever your heart trusts in and relies on, that is properly your God." (Ramsey 1970, 143–144)

In sum, the TH claim is pervasive in public policy bioethics, and it resurfaces in the synthetic biology debate. Usually the claim is not opposed, but simply ignored, because for the reasons described above it cannot be adjudicated in mainstream bioethics.

A Method for Evaluating whether a Teaching Humanness Effect Exists

I think that the mainstream of bioethics has ignored the importance of empirical examination of the public's values, and I have recommended elsewhere a plan for the incorporation of the public's values into public policy bioethics (Evans 2012). Most of the bioethics arguments I read rely on implicit empirical assumptions about how humans will behave, and if ethicists teamed with social scientists who could show which of those assumptions are plausible, a much better ethical product would result. This type of partnership is the mild social-science intervention into bioethics that Hedgecoe, following Nelson, calls the "linear model" (Hedgecoe 2007, 167). These TH claims really are an instance where social science can help clear the brush for ethicists. While in my more detailed proposal I call for empirical observation of the public's values (Evans 2012), here I offer a simpler sketch that suggests how we could proceed.

Of course we should not act on a TH claim unless the claim is legitimate. The first step in evaluating these claims would be to take the notion of humanness that the technology is claimed to cause, and see if the public already holds that notion. If so, then all other questions become moot and no further investigation is needed. For instance, if the public already thinks of itself as gods, this new technology will not produce that view.

To determine if the public already holds the notion that humans are creators, one could conduct an empirical study by basically asking them. This is standard fare in the field known as cultural sociology, where studies have shown, for example, that certain types of individualism are dominant in American culture (Bellah et al. 1985). If people already think of humans as creators in the specific sense supposedly taught by synthetic biology, then analysis can stop and the claim is dismissed.

If a claim has passed the first step, the next step is to determine whether the technology could actually cause the change. One method would be to run an experiment where you exposed a group from the public to the technology in question and see if they developed the particular notion of the human. Something very similar has been done to assess whether mass-media coverage of genetics leads to increased belief in genetic determinism (Condit 1999). One could also infer that if the people who are disproportionately exposed to the technology, like synthetic biology scientists, are more likely to have the notion of the human in question, then there is at least an affinity between the technology and the notion of the human, if not a causal link. With these two steps completed, we could conclude that the TH claim is legitimate, and the transformation in our sense of our humanness should be opposed. Even if we do not conduct these social science analyses, by bringing in existing data about society we can identify which claims deserve further investigation, as I do below.

This method is actually a specific instance of what Erik Parens calls the "argument from precedent" (Parens 1995). It is extremely common to argue that "technology X is just like Y, and we already accept Y, so we should accept X." For example, Bedau and Triant make a classic argument from precedent to refute the claim that synthetic biology is "unnatural":

> One might consider it "unnatural" to intervene in the workings of other life forms. But then the unnatural is not in general wrong; far from it. For example, it is surely not immoral to hybridize vegetable species or to engage in animal husbandry. And the stricture against interfering in life forms does not arise particularly regarding humans, for vaccinating one's children is not generally thought to be wrong. So there is no evident sense of "unnatural" in which artificial cells are unnatural and the unnatural is intrinsically wrong. (Bedau and Triant 2009, 33)

The argument from precedent uses a seemingly universally accepted example to stand in for public opinion (for example, people accept mules), whereas my method uses a more direct measure. Moreover, these arguments always seem to assume that the public is logical and consistent, and that what features of the example are salient to the analyst will also be salient to the public. For example, in the above quote, it is assumed that the public considers vaccinations to be an equivalent "interference" to genetic modification for a human, which is not likely (see below).

Presentist Orientation

The proposed method of testing for a TH effect is based upon measuring the dominant conception of the human in the present and comparing it to an empirically informed yet hypothetical future. We could imagine methods whereby scholars try to show larger gaps by reaching back in time to an era where a much different notion of the human reigned. One could study pre-Darwin novels, argue that they represent a society with a different dominant assumed humanness, and then compare that conception to the hypothetical humanness of 2050, when technology X will be pervasive. The technology would then be bad because it is partly responsible for a change in notions of the human that is disapproved of.

While I would not object to this being done for rhetorical purposes—to be able to demonstrate how far we have slid down the slippery slope—I think that only the preferences of present-day people should be taken into account, even if these people have already been partially dehumanized (in, for example, Kass's view), by accepting bad notions of humanness taught by previous technologies. If on a scale from 0 to 100—with the end points being a Christian notion of the human supposedly held in, say, the seventeenth century, and a biological humanness—if we are at 50, then people will probably not mind moving to 51.

It is the current public's views that should be determinative. As already noted, most of public policy bioethics accepts that ethical claims that influence policy must be based on the current public's ethics (even if bioethicists see their role as clarifying and specifying those ethics.) For example, theories of common morality are exactly this—an attempt to say that any policy recommendation is simply a crystallization of the morality that is already held by the American people (Evans 2012). Statements of opposition to TH targeted to the public should then be based on whatever theories the speaker wishes, while statements of opposition geared toward actual policy should be based on what the public thinks. Therefore, if a TH effect is convincingly demonstrated, a final step before acting would be to determine whether the public would actually mind the change in conception of the human.

Many proponents of TH arguments will be very disappointed in one implication of making the values of the current public determinative, which is that small changes in our notions of humanness will not be detectable by empirical methods, and even if they are, they will not be of a magnitude to garner political interest or even justify acting. Therefore, the slow, steady erosion of our humanity, in some views, cannot be opposed. This is quite a sacrifice for proponents.

I still remember Gilbert Meilaender's testimony to the National Bioethics Advisory Commission in 1997 regarding what he saw as the dehumanization caused by reproductive cloning. Challenged by the argument from precedent (Parens 1995), he was asked whether other existing reproductive technologies like IVF caused the same dehumanization problems, and if so, should he then not oppose this technology. Meilaender simply said that yes, earlier technologies did, and he would have gotten off the train of transformation in human consciousness long ago, presumably before the advent of IVF in the 1970s. That is, rarely is the claimed change in our humanness qualitatively different from that which preceded it, nor is it often a change of any great magnitude. The claim typically entails that the technology is another small step down the slippery slope toward the undesirable end. Meilaender can see a huge transformation in our notion of humanness by imagining himself back to the past. But the public has, in Meilaender's eyes, already been dehumanized to some extent, so further dehumanization does not look so different to the observer in the present. (Kass's similar statement was quoted in the previous section.)

For example, in the "teaching a divine humanness" version of the TH claims about synthetic biology, imagine the continuum of thinking of ourselves as created versus creator. The first slide down the slippery slope probably began with human fabrication of tools, down to the Industrial Revolution, through to human interventions in reproduction and the engineering of plants, animals, and microbes. We probably think of ourselves as substantially more godlike than people did 400 years ago. If people who make the TH claim about synthetic biology are correct, we are at present staring down our slope at a possible future where we learn that we are yet more godlike.

The democratic version of ethics that I am appealing to—where the ethics of the present's people should determine the future—means that it does not matter that this last change would be momentous compared to 100 years ago. Moreover, measuring a gap from the present to the future, instead of 100 years ago to the future, will make it difficult to find an effect of a technology. Rather, if synthetic biology would be an incremental change, then people would not be opposed to it, because it would not look like a change. We could then only oppose large jumps down the slippery slope. If synthetic biology were being proposed in 1965, before we gained the ability to be what Boldt and Müller call "homo faber" when it comes to bacteria, combining genes from different life-forms, then the shift in human self-perception would indeed be great. It would be the sort of change that probably would have been opposed. However, at present, as I will argue below, if it exists at all, the transformation in human self-conception is not large enough to be observed.

Objection to Imprecise Methods

Besides complaining that I have just described the sociologist full employment act, proponents of synthetic biology will say that we should never change policy toward a promising technology like synthetic biology based on the results of studies like these, which will undoubtedly be very fuzzy by the standards of the natural sciences. This complaint can be dismissed by the adage about the drunk looking for his lost keys under the lamppost because that is where the light is. You cannot ask only the easy ethical questions. The question should be, Is this a better way of evaluating TH claims than we currently have? I would argue that the current method is armchair reflection by the analyst on what the public thinks and how the public's beliefs would change due to a technology. How does Leon Kass know that we don't already think of ourselves as pork-bellies? The method discussed here would have to be better, if only in that it requires those making claims to state what they believe about what the public would think and so on.

Moreover, our society currently sets policy by evaluating the values of the public through far less precise methods. For example, the dominant form of argumentation used in bioethics, especially in research bioethics and health care ethics consultation, is based on the four universal values of Western civilization embedded in principlism. Yet these were determined, through scholarly reflection, at a meeting of a dozen or so academics (Jonsen 2005). Moreover, the structural mechanism for the expression of our values in public affairs is our elected officials, who are supposed to represent the values of their constituents. It would take a large leap of faith to describe the electoral system as accurate in this regard. In some states we decide policy through ballot initiative. For example, in California the citizens enacted a policy of funding embryonic stem cell research a few years ago. Not only did that policy take effect despite the 41 percent who voted against it, but given the way the initiative debates were conducted, it is hard to argue that the vote was an accurate measure of Californian values.

Of course, a requisite degree of caution should be used in reaching conclusions from studies such as I am proposing. Even if we were to take such claims with a dose of skepticism—and we would want a number of different scholars trying to evaluate the same claims—I have not seen any other suggestions for evaluating TH claims, beyond either ignoring them or simply asserting what I am trying, instead, to measure.

Objections to Acting on Teaching Humanness Claims in Principle

Beyond concerns about the method, some people may be opposed in principle to acting on a TH claim even if it is shown to be legitimate. A possible objection is that we have no business trying to influence what people in the future will think about themselves. However, much of our ethics is based upon trying to create a particular sort of world for future persons—think of the environmental movement. Moreover, the debate about the U.S. government torturing people is often put in terms of "what kind of country we want to be"—that is, present people are trying to create a particular country for future Americans. We present people are obviously the most powerful force determining what future people will think of themselves, so we might as well be conscious about what we are doing.

Some bioethicists would say that all of the claims that I am calling "thick" are not legitimate in a pluralistic liberal democratic society, because in such a society we do not set the values for people, but rather let people pursue their own values. The proponents of limiting ethical claims to what I am calling the "thin," or to those claims that can be translated to the "thin," are motivated by a very strong Rawlsian political philosophy in which only values with a high degree of overlapping consensus can be acted upon by the liberal democratic state. I would argue that this sets an impossible standard that serves to simply promote a bias toward the value of autonomy on the part of people advocating this view, because if we cannot act collectively on values, then autonomy reigns. There is no public policy that is derived from value consensus in society, or even near-consensus. We have elections because we disagree, and our elected officials therefore make decisions based on less than consensual values. So, while I am in favor of basing policy on the values of society, we can talk about more values than the four supposedly universal values expressed in principlism. Of course, if one can make the case that engaging in the technology is constitutionally protected, or that the values are actually "religious," and thus their use is a violation of the First Amendment, then this procedure would not be used.

The alternative to trying to set the values of the future in at least as consensual a way as possible is not to act at all. In actuality, the collective self-conception of our descendants will be shaped by people acting in the present whether we want it to happen or not. Put more bluntly, if we do not try to influence the values of the future, we are letting contemporary people with power and money do it for us.

An Initial Evaluation of Synthetic Biology "Teaching Humanness" Claims

I have conducted a preliminary analysis of TH claims in synthetic biology using existing social science data, and have found these claims at present unjustified. A more complete examination would require a dedicated research project, but by walking through this example I hope to sharpen the questions we should ask. Starting with the "teaching a divine humanness" version, what Boldt and Müller are arguing is that contemporary humans already see ourselves as *fabricators* of life-forms—as we have done through centuries of breeding tangelos and mules, and more recently, genetically engineered microbes—but not as *creators* of life de novo. This notion of humans as creators is new, and this is what they oppose. Even more precisely, the claim is that there is a cognitive domain titled "human" (and below I provide evidence that such a domain exists), and that the features of "the human" would now include "creator of life," displacing the feature "manipulator of life."

However, at present, I think that empirical evidence shows that people will not recognize the distinction between fabricators and creators, so being taught that they are creators is not going to change their self-perception. We can already accept that part of being human is to be a creator, not only a fabricator. Thus this claim fails at the first step.

In a separate project I interviewed 180 members of religious congregations across the United States who were broadly representative of the religious population as a whole—liberal, conservative, and otherwise—concerning what they thought about human reproductive genetic technologies (Evans 2010). I examined what people thought of the relationship between God, nature, and humanity, and my conclusion was that how humans would achieve a goal does not register to them, but rather they are focused on what the goal is. Certain goals *are* considered outside of the nature of the human. This was shown by the fact that the largest group among these church members did not have a problem with humans having any technological ability. This idea was often described as the problem of knowing what God wants us to do with our wonderful technological abilities. For example, Beatrice, a traditional evangelical, says:

> It goes back to the sense that God has given us these abilities and this intelligence to create penicillin to wipe out, not wipe out, but to able to fight bacterial diseases, to create drugs that combat cancer and other disease. It gives me medicine to take for my thyroid to make it work right, and I think that is all good stuff. I think there can be good stuff coming out of genetics and genetic technology. But there is always the flip side of how that technology is going to be used. I mean, we create all these drugs, and we also create dangerous viruses that can be used as weapons. That is the flip side of everything. God has given us the ability, so I can't . . . I don't believe you can say, "Okay, God alone can do this," but on the other hand I believe you have to be willing to look at things as not just as "What can I do?" but "What should I do? What is the right thing to do?" Unfortunately, too many people don't care about what is right. They just want to do whatever they want to do because they can do it. I think that is a problem.

Alice, another traditional evangelical, explained her similar view through a story that anyone who spent a lifetime in a church would eventually hear:

> There was a man in his house during a torrential rain, and the firemen came by and told him to leave for his own safety. And the man says, "No, I have faith in God." And so he goes up on the roof as the water rises and the police come and ask him to leave and everything. And the man says, "No, I have faith in God, God will save me." Then a sailboat came by, they tell him to step in the boat, you know that it's flooded, you know, everybody's evacuated and he says, "No, I have faith in God, I'm staying and I'm not leaving my house." . . . Pretty soon the helicopter comes and they let down a rope for him to grab on, you know, that if he doesn't get off that roof soon the house is just gonna wash away and he's gonna die. He

says, "No, I have faith in God. God's gonna take care of me and I'm gonna stay right here." So the man dies and he gets to heaven and he asks the Lord, "You know, I had faith in You and—and You didn't save me. Why didn't You save me?" And He said, "I sent the firemen. I sent the helicopters . . . ." You know, so it's kind of like that, you know. I mean, His hands are our hands, you know. And so if He has sent things that we can use to help us then, yeah. Go for it.

A related theme is expressed by Jack, a middle-aged fundamentalist who has a more interventionist God in mind. It is not so much that God gave us the brains to figure things out, but rather that God allows us in each instance to develop a technology. He told us that "certainly we've been given technology by God to use. A lot of times it gets misused but I believe the technology is there because God allows us to have it. Why? I don't know exactly and why did he wait so long to decide to give us some of the technology that we do have? I don't know. . . . I mean, using the analogy that I've used, it's just like a medical procedure that would eradicate any other kind of disease." A young mainline woman said something similar when asked whether the ability to change the genes of the human species should be reserved for God. "God has given us the ability to do it," she began. "So, apparently, God doesn't think it should be reserved for God 'cause He's already letting us do it."

Of course, there were a group of people who thought that God had put certain powers out of the reach of humans. These people were the strongest opponents of reproductive genetic technologies, but, critically for my point here, they were not the largest group. While I did not specifically focus upon the nonreligious citizens, from what we know about American society, I would not expect them to take the noninterventionist view.

Another striking feature of these interviews was that the technology used to achieve genetic goals does not matter to people. At the core of the interview, I asked about the use of amniocentesis to ensure that a child did not have cystic fibrosis. I then asked about the use of the same technology for gender, and then for enhancements. I then turned to preimplantation genetic diagnosis, and asked again about the use of this technology to avoid cystic fibrosis, then gender, then various other conditions hovering between disease and enhancement like deafness, and then asking about using the technology to enhance a child's intelligence. I then started over again with questions about human genetic engineering.

The structure of the responses is striking. You get a very similar pattern of responses to the *goal* of the technology, regardless of the technological means used to achieve it. I had thought that the interview guide would identify these differences between technologies, but the overall conclusion is that it is the *outcome* of the technology that matters for people. People who approved of amniocentesis for cystic fibrosis tended to approve of every other way to avoid cystic fibrosis. People who approved of using abortion for gender selection tended to approve of any other means of achieving gender selection. How we achieve our goals does not matter to contemporary people, at least with genetic technologies, and what matters is what we are trying to do with the technologies. The primary moral divide was between "disease" and "enhancement," no matter where people drew that line. In theological terms, people thought that God had already assigned to humans (i.e., it is an aspect of the human) the task of healing disease, but had not assigned to humans the task of inventing themselves into new creatures with new characteristics.

This means that the technology itself does not register. It is already within the definition of the human to heal disease *by any means or any technology*. However, it is not within the definition of the human to invent new human qualities—that is, the characteristic "creates new human features for the human" is not a feature of the cognitive domain of "human." In fact, to use the terminology of this paper, I had thought that people would oppose human genetic engineering (described to the respondent as changing the genetic characteristics of all of your descendants) on the ground that it is something that humans do not do, that only God does. I was wrong.

So, the lesson from this is that the ability of synthetic biology to create life itself for ends that we have already assigned to ourselves as humans, like reducing pollution or developing energy, will not be seen as a new ability of the human. So, most of what is proposed for synthetic biology would not "teach this divine humanness." I would argue that if creating new life-forms to fulfill goals that we already think of as part of the human project is tantamount to considering ourselves gods, then we already think of ourselves as gods. This TH claim fails the test.

The "teaching a biological humanness" version of the synthetic biology claim would probably pass the first step of the analysis. Although I do not have data on this question yet, I think it is reasonable to presume that humans do not presently think of themselves as biological beings, but as something like spiritual beings. I do not think that many people think of themselves as defined by their DNA. So, this would be a change in the conception of humanity of the magnitude that could be opposed.

However, I think that if I applied the third step of the evaluation, we could debate whether the technology in question could cause the change in our notion of humanity. The claim is that if people come to think of bacterial life as simply DNA, then they will come to think of human life as simply DNA. This assumes that people make an analogy: if "life" of a bacteria is like X, then "life" of a human is also like X, because they are both life-forms. On the one hand, anthropologists who study folk biology examine how diverse societies classify different life-forms. The Tzeltal produce a "classification system which partitioned the world of living things into 'humans,' 'animals,' and the polylexemic phrase 'trees and plants'" (D'Andrade 1995, 94). But it is not only the Tzeltal. In fact, these distinctions are nearly universal, with other scholars describing the divide as person, artifact, plant, or animal (de Cruz and de Smedt 2006, 352). They are so universal that evolutionary psychologists have seized on the idea that these basic distinctions were adaptive, and therefore the basic biological structure of our brains is designed to make these particular distinctions. Evidence that this divide is biological is also found from studies of people with various types of brain damage, like the woman who cannot recognize anything having to do with animals, but her understanding of plants and artifacts remains unimpaired, suggesting that these distinct domains are in different locations in the brain (de Cruz and de Smedt 2006, 353).

Whether we think that the near universality in these distinctions is based in biology or on universal features of human experience, it seems clear that this trait would work against making category mistakes between humans, animals, plants, and inanimate objects. Put differently, that the defining characteristics of the "life" of a bacteria may be purely biological is not likely to lead people through analogical reasoning to conclude that the defining characteristics of the "life" of humans are also biological.

Further evidence that people make these categorical distinctions comes from the teaching humanness claim itself. People who make this claim say that they are worried that defining bacteria by their DNA will lead humans to reject the idea that human life is "special." This observation of the "special" status of human life is evidence for the argument that people currently consider human life to be different and that they will not make this category mistake.

On the other hand, we cannot close the door on the possibility of the category mistake. History is full of examples of concepts moving across the boundaries, such as concepts of nature influencing concepts of society, and of views of animals influencing views of humans.[1] For example, the idea of "inhibition" in human behavior was related to ideas of inhibition in machinery (Smith 1992). While we do not have a definitive conclusion about whether people will make this category mistake, we at least we know the sort of question we should be asking.

What Should Be the Response to Teaching Humanness Claims?

If the TH claim is found to be legitimate, the response should be to restrict what actually causes the transformation in our sense of humanness. How then does a person come to believe that humans are like X? I believe that I am in the sociological mainstream when I say that people are taught through discourse that humans are like X, but this discourse is influenced by actual experience.

For people to accept a discourse about the nature of humans that is discordant with what they experience humans to really be like requires more consistent exposure to the discourse. This is evident from what we know about the unusual religious groups usually labeled "cults." If you lock people in a compound for ten years and tell them constantly that they are beings with superpowers from another planet, they will believe this so strongly that they will, and have, killed themselves because of this understanding, believing that their actual essence will be transported to an awaiting spaceship. Or, in a case of less complete domination by discourse, Nazi propaganda and other sources of discourse convinced Germans that Jews, Roma, and others were less human than others and therefore killing them was not a problem. Discourse can overwhelm what we consider to be the reality of experience if there is enough discourse.

Therefore, teaching humanness happens through implicit and explicit discourse that essentially says, "because we can synthesize viruses we humans are creators, not fabricators," as well as by people experiencing themselves as creators through the use of synthetic biology technology and coming to the conclusion, without being told, that humans are therefore creators. Our response to a threat of changing our humanness can then be either to use a discourse to describe the technology that will not result in the change, or to eliminate the experience of the technology by banning it.

The first of these solutions sounds like lying about the true nature of a technology. However, it is interesting that proponents of TH claims do not seem to believe that the experience of the *true* nature of the technology *should* lead to the transformation. Meilaender, Kass, and Boldt and Müller do not think of themselves as more godlike due to their knowledge of technology, but seem to instead be fearful that others will make the mistake. For example, Boldt and Müller write that we will "come to understand ourselves, *justified or not*, as creators" (2008, 388; my emphasis). I suspect they are immune from the effect of the experience of the technology because they have a more sophisticated understanding of either the technology or the human. From the viewpoint of those opposing TH claims, then, what is required is not a lie, but the proper and more precise discourse to describe the technology, which would say, for example, about synthetic biology, "While it may appear that you are a creator, in actuality you are not."

The mildest form of intervention would be to try to convince the primary discourse producers to voluntarily stop describing the technology in a way that would cause the change in human self-conception. It has been recognized that promoters of a technology have to talk in grandiose ways in order to get attention and funding for their technology (Yearley 2009), but they could be made aware that talking in this way may damage our sense of humanness. Or, if they don't care about a change in our self-conception, they could be made aware that talking like they have been talking may cause a political backlash. They would not have to describe the technology inaccurately.

It seems clear that in an earlier debate about human genetic engineering, if scientists had not made public claims that the new technology could redesign the human species and produce meaning and purpose for all of humanity, they would not have drawn the same attention, not only from the theologians who came to oppose them, but also from people like Jeremy Rifkin, who would be a thorn in their side for decades (Evans 2002). For example, in 1969, then proponent of germline genetic engineering Robert L. Sinsheimer wrote that the emerging genetic engineering technologies allowed for a "new eugenics." "The old eugenics would have required a continual selection for breeding of the fit, and a culling of the unfit. The new eugenics would permit in principle the conversion of all of the unfit to the highest genetic level . . . for we should have the potential to create new genes and new qualities yet undreamed" in the human species. Even more, he described the new technologies as "the turning point in the whole evolution of life. For the first time in all time, a living creature understands its origin and can undertake to design its future. Even in the ancient myths, man was constrained by his essence. He could not rise above his nature to chart his destiny. Today we can envision that chance—and its dark companion of awesome choice and responsibility" (Sinsheimer 1969, 13). Not talking this way about human genetic engineering circa 1969 would not have been lying. As at least one scientist (correctly) surmised at the time that the "irresponsible hyperbole" of scientists had negatively influenced the funding of research. "Some of these statements, and many articles in the popular press, have tended toward exuberant, Promethean predictions of unlimited control and have led the public to expect the blue-printing of human personalities," he wrote. The "exaggeration of the dangers from genetics will inevitably contribute to an already distorted public view, which increasingly blames science for our problems and ignores its contributions to our welfare" (Davis 1970, 1279).

Either scientists have learned nothing since 1969, or the need to make grandiose statements in order to obtain funding for research has remained a constant, because the promoters of synthetic biology sound very similar to Sinsheimer. When a colleague of the synthetic biology pioneer Craig Venter was asked by an interviewer whether they are "playing God," he reportedly responded, "We don't play" (ETC Group 2007, 15). More mildly, advocates seem intent on shocking people into the conclusion that their research is important by the extensive use of analogies to inanimate objects. The primary analogy is of the minimal genome "chassis" upon which "cassettes" are placed. There is also the "hardware" of the minimal genome bacteria and the "software" of the traits we want to engineer. One scientist said, "We're building the modern chemical factories of the future" (ETC Group 2007, 19). While I am not positive that convincing people that bacterial "life" is analogous to an inert object will ultimately damage our sense of our humanness, if I were, I would say that scientists should stop talking like this, especially since it's not actually necessary to describe their technology.

Government policy could also be used to fight TH effects. First, consider the discursive part of the effect. Governments will not ban discourse, but they regularly promote certain discourses. Since I am not yet convinced of the TH claim for synthetic biology, let me use one that strikes me as more clear, that pricing babies will lead to our thinking of people as not ultimately valuable—"cheapening the value of life." I think current government policy is to allow the experience of pricing babies, but to create a discourse that masks the pricing, thus avoiding any transformation in human conception about babies. The government is essentially producing a counterdiscourse to what our experience might lead us to conclude, and, whether intentionally or not, it thereby limits the change in how we see ourselves.

To anyone who has contemplated adopting a baby, it is not news that they cost different amounts of money. This money is all labeled as "fees" of various sorts, but white babies effectively cost more than black babies, which effectively cost more than black babies that have been taken away from their parents by the state. Something similar can be said about how producing a baby from "donated" eggs from a woman currently enrolled in an Ivy League institution costs more than the baby produced by eggs from "ordinary" women. The key is that while all this is allowed by the government, it is described to the public as fees for work, not as prices that result from differential value. This sort of discursive practice with regard to synthetic biology will go far toward avoiding a change in our human self-conception, especially if any counterdiscourse about what is "really" happening can be discouraged.

Of course, reality can overwhelm the effects of discourse. Like the cult member who is allowed to go to the grocery store and then realizes she doesn't actually have superpowers, if people actually experience the technology, they may conclude that the discourse surrounding it is false. This can work for both good and bad: people will meet a clone and realize they are just as human as everyone else, or experience synthetic biology and conclude, despite being told it does not, that it does make us gods. So, a stronger response would be to ban the experience, so that nobody comes to realize the truth of the matter. With pricing babies, we could ban the experience of differential pricing of persons by charging a universal baby fee for any adoption, and using the "profits" from the cheaper processes to subsidize the more expensive processes. If society really concludes that people who learn of synthetic biology will conclude that they are gods, no matter how it might be described otherwise, then society should ban synthetic biology.

Whether it would be worth it to eliminate the experience that causes a lack of faith in the preferred conception of humanness depends upon two variables. The first is the number of people who would actually have the experience. The experience of pricing babies is fairly widespread, but not as widespread as would be the effect of putting a price on being kept alive at the end of one's life, if we enacted such a policy. The actual experience that causes the TH effect for synthetic biology is the act of creation, which very few people would be privy to, compared to the consumption of synthetic biology products, which should not have the effect. With few people having the experience, there is less reason to ban it. The other variable is the size of the gap between what people are going to conclude from the experience and how the experience is described in discourse. To put it more crassly, if there is no way to "spin" an experience, then advocates have no choice but to ban it. However, most technological experiences have a number of alternative interpretations.

Conclusion

Teaching humanness claims are sociologically plausible, but would need to be evaluated for each case. They are the "thickest" of the "thick" claims in public policy bioethics because they are explicitly about setting our collective values, not defending institutionalized and assumed-to-be universal values. Such claims are a legitimate basis for public action, and the criteria for whether or not to act should be whether the transformation would be opposed by presently existing persons. While I think that TH claims are in general plausible, my preliminary analysis of the claims made in the emerging synthetic biology debate suggests that one of these particular claims is not, and the other is unclear. In the face of a plausible TH claim, people should engage in some combination of changing the discourse surrounding the technology or limiting the number of people who experience the technology.

Beyond a lack of clarity or wisdom on my part, the following are reasons why my argument will be ignored. First, as noted above, public policy bioethics does not want to engage in "thick" arguments because they are not actionable through bureaucratic government response. The idea that some bureaucrats are determining what our values should be is anathema in the American (but not the European) context. Second, to my knowledge, no one has ever specified a "dehumanization" claim—often made in support of banning technologies like reproductive cloning—down to the level where some claims could be determined to be right and others wrong. I think some people will resist specifying the claim. Third, people who want their ethics to be used by the entire society regardless of whether the public agrees will not like the fact that the views of the contemporary public are the standard for determining whether an ethical problem exists. In particular, advocates who know that the technology will result in yet another incremental slip down the anthropological slippery slope will not like the fact that they cannot refer back to how our parents would have viewed the world. Fourth, many will not like the idea of restricting technologies on the basis of empirical research about what contemporary citizens will think, because such research will inevitably be more fuzzy and less precise than the (unfounded) assertions in the bioethical literature about what the public thinks. Of course, if anyone actually conducted a research project to determine the actual meaning of the "autonomy" that is purportedly a universal moral feature of Western civilization, it would make all bioethics, and especially the parts of bioethics that rely upon the bureaucratic state, such as institutional review boards, unmanageable.

I think it is inaccurate to simply assert what the public thinks, so we should endeavor to find out. Yes, it will be messy, and complicated. But nobody says, "Resolving global warming is messy and complicated, so we shouldn't try to do it." We should instead press on by doing our best. If these claims are going to be taken seriously, people who make them in the future should provide preliminary evidence that their claim would survive the three stages of this analysis. For example, authors should provide evidence that humans will transfer the concept of "life" across domains.

Biological innovation is moving fast. There will be an increasing number of teaching humanness claims. People may snicker at the transhumanists who, for example, want their offspring to have wings, but those who do not want that to happen will have only a few ways of opposing their "reproductive freedom." People will be able to argue that it is not safe. But if it is safe, then the argument will be that it creates a transformation in human self-conception that we disapprove of. Before the first transhumanist baby soars from its nest, we need to find a way to see whether such objections are truly sound and worth acting upon.

Acknowledgments

This paper was presented to audiences at the Conference on Ethical Issues in Synthetic Biology at The Hastings Center, the ESRC Centre for Economic and Social Aspects of Genomics at the Universities of Cardiff and Lancaster, and the Cultural Studies of Science and Technology Seminar at Rice University. Thanks specifically to Steve Sturdy and Greg Kaebnick for comments.

Note

1. Thanks to Steve Sturdy for making this point.

References

Beauchamp, Tom L. 2003. A defense of the common morality. *Kennedy Institute of Ethics Journal* 13 (3):259–274.

Bedau, Mark A., and Emily C. Parke. 2009. Social and ethical issues concerning protocells. In *Protocells: Bridging Nonliving and Living Matter*, ed. S. Rasmussen, M. A. Bedau, L. Chen, D. Deamer, D. C. Krakauer, N. H. Packard, and P. F. Stadler, 641–643. Cambridge, MA: MIT Press.

Bedau, Mark A., and Mark Triant. 2009. Social and ethical implications of creating artificial cells. In *The Ethics of Protocells: Moral and Social Implications of*

Creating Life in the Laboratory, ed. M. A. Bedau and E. C. Parke, 31–48. Cambridge, MA: MIT Press.

Bellah, Robert N., Richard Madsen, William M. Sullivan, Ann Swidler, and Steven M. Tipton. 1985. *Habits of the Heart: Individualism and Commitment in American Life*. New York: Harper and Row.

Boldt, Joachim, and Oliver Müller. 2008. Newtons of the leaves of grass. *Nature Biotechnology* 26 (4):387–389.

Burtt, Edwin Arthur. 1959. *The Metaphysical Foundations of Modern Science: A Historical and Critical Essay*. London: Routledge and Kegan Paul.

Caplan, Arthur L. 1997. *Am I My Brother's Keeper? The Ethical Frontiers of Biomedicine*. Bloomington: Indiana University Press.

Cho, Mildred K., David Magnus, Arthur L. Caplan, Daniel McGee, and the Ethics of Genomics Group. 1999. Ethical considerations in synthesizing a minimal genome. *Science* 286:2087–2090.

Condit, Celeste M. 1999. How the public understands genetics: Non-deterministic and Non-discriminatory interpretations of the "blueprint" metaphor. *Public Understanding of Science* 8:169–180.

Condit, Celeste M., Nneka Ofulue, and Kristine M. Sheedy. 1998. Determinism and mass-media portrayals of genetics. *American Journal of Human Genetics* 62:979–984.

D'Andrade, Roy. 1995. *The Development of Cognitive Anthropology*. New York: Cambridge University Press.

Davis, Bernard D. 1970. Prospects for genetic intervention in man. *Science* 170:1279–1283.

De Cruz, Helen, and Johan de Smedt. 2006. The role of intuitive ontologies in scientific understanding: The case of human evolution. *Biology and Philosophy* 22:351–368.

Discovery Institute. N.d. The Wedge. http://www.antievolution.org/features/wedge.pdf.

ETC Group. 2007. *Extreme Genetic Engineering: An Introduction to Synthetic Biology*. Online publication of the ETC Group. http://www.etcgroup.org/content/extreme-genetic-engineering-introduction-synthetic-biology.

Evans, John H. 2002. *Playing God? Human Genetic Engineering and the Rationalization of Public Bioethical Debate*. Chicago: University of Chicago Press.

Evans, John H. 2010. *Contested Reproduction: Genetic Technologies, Religion, and Public Debate*. Chicago: University of Chicago Press.

Evans, John H. 2012. *The History and Future of Bioethics: A Sociological View*. New York: Oxford University Press.

Garfinkel, Michele S., Drew Endy, Gerald L. Epstein, and Robert M. Friedman. 2007. *Synthetic Genomics: Options for Governance*. Rockville, MD: J. Craig Venter Institute.

Gustafson, James M. 1990. Moral discourse about medicine: A variety of forms. *Journal of Medicine and Philosophy* 15:125–142.

Hedgecoe, Adam. 2007. Medical sociology and the redundancy of empirical ethics. In *Principles of Health Care Ethics*, 2nd edition, ed. R. E. Ashcroft, A. Dawson, H. Draper, and J. R. McMillan, 167–175. West Sussex, UK: John Wiley and Sons.

Jonsen, Albert R. 2005. On the origins and future of the Belmont Report. In *Belmont Revisited: Ethical Principles for Research with Human Subjects*, ed. J. F. Childress, E. M. Meslin, and H. T. Shapiro, 3–11. Washington, DC: Georgetown University Press.

Kass, Leon R. 1985. *Toward a More Natural Science: Biology and Human Affairs*. New York: Free Press.

Kass, Leon R. 1992. Organs for sale? Propriety, property, and the price of progress. *Public Interest* 107 (Spring):65–86.

Kaye, Howard L. 1997. *The Social Meaning of Modern Biology: From Social Darwinism to Sociobiology*. New Brunswick, NJ: Transaction Publishers.

Kaye, Howard L. 1998. Anxiety and genetic manipulation: A sociological view. *Perspectives in Biology and Medicine* 41 (4):483–490.

Meilaender, Gilbert. 1997. Begetting and cloning. *First Things* (June/July): 41–43.

Nelkin, Dorothy, and M. Susan Lindee. 1995. *The DNA Mystique*. New York: W. H. Freeman.

Numbers, Ronald L. 1982. Creationism in 20th-century America. *Science* 218:538–544.

Parens, Erik. 1995. Should we hold the (germ) line? *Journal of Law, Medicine, and Ethics* 23:173–176.

Parens, Erik, Josephine Johnston, and Jacob Moses. 2008. Do we need "synthetic bioethics"? *Science* 321:1449.

Ramsey, Paul. 1970. *Fabricated Man: The Ethics of Genetic Control*. New Haven: Yale University Press.

Rothman, Barbara Katz. 1989. *Recreating Motherhood: Ideology and Technology in a Patriarchal Society*. New York: Norton.

Schummer, Joachim. 2009. The creation of life in cultural context: From spontaneous generation to synthetic biology. In *The Ethics of Protocells: Moral and Social Implications of Creating Life in the Laboratory*, ed. M. A. Bedau and E. C. Parke, 125–142. Cambridge, MA: MIT Press.

Sinsheimer, Robert L. 1969. The prospect for designed genetic change. *American Scientist* 57 (1):134–142.

Smith, Roger. 1992. The meaning of "inhibition" and the discourse of order. *Science in Context* 5 (2):237–263.

Turner, Leigh. 2003. Zones of consensus and zones of conflict: Questioning the "common morality" presumption in bioethics. *Kennedy Institute of Ethics Journal* 13 (3):193–218.

Van Est, Rinie, Huib de Vriend, and Bart Walhout. 2007. *Constructing Life: The World of Synthetic Biology*. The Hague: Rathenau Instituut.

Yearley, Steven. 2009. The ethical landscape: Identifying the right way to think about the ethical and societal aspects of synthetic biology research and products. *Journal of the Royal Society, Interface* 6:S559–S564.

About the Authors

John Basl is an assistant professor of philosophy at Bowling Green State University. He is also a cofounder and cocreator of Philosophy TV (http://philostv.com).

Mark A. Bedau is professor of philosophy and humanities at Reed College, editor-in-chief of the journal *Artificial Life*, and a regular visiting professor in the Foundations and Ethical Implications of the Life Sciences program at the European School of Molecular Medicine in Milan, Italy. He is coeditor of the recent books *Emergence* (2008), *The Nature of Life* (2010), *Protocells: Bridging Nonliving and Living Matter* (2008), and *The Ethics of Protocells: Moral and Social Implications of Creating Life in the Laboratory* (2009).

Joachim Boldt is assistant professor in the Department of Medical Ethics and the History of Medicine and associate member of the German Research Foundation's Cluster of Excellence BIOSS (Biological Signalling Studies) at Freiburg University, Germany. He is editor, with Oliver Müller and Giovanni Maio, of *Leben schaffen? Philosophische und ethische Reflexionen zur Synthetischen Biologie.*

John H. Evans is professor of sociology at the University of California, San Diego. He is the author of *Contested Reproduction: Genetic Technologies, Religion, and Public Debate* (2010) and *The History and Future of Bioethics: A Sociological View* (2012).

Bruce Jennings is director of bioethics at the Center for Humans and Nature and senior advisor at The Hastings Center. He is coeditor, most recently, of *Public Health Ethics: Theory, Policy, and Practice* (2007) and coauthor of *The Perversion of Autonomy: The Proper Uses of Coercion and Constraints in a Liberal Society* (2003).

Gregory E. Kaebnick has been a coinvestigator on two research projects at The Hastings Center on synthetic biology. He is the editor of *The Ideal of Nature: Debates about Biotechnology and the Environment* (2011). He is also the editor of the *Hastings Center Report*.

Ben T. Larson graduated from Reed College with a degree in physics. He is currently working on molecular and cellular biophysics at the NIH.

Andrew Lustig is the inaugural holder of the Holmes Rolston III Chair in Religion and Science at Davidson College in North Carolina. He has published widely in bioethics and theology. His most recent books are the two-volume set *Altering Nature* (2010).

Jon Mandle is a professor in the Department of Philosophy at the University at Albany (SUNY). He is the author of articles and books in political philosophy, focusing especially on the work of John Rawls.

Thomas H. Murray is president emeritus and a senior research scholar at The Hastings Center. He has been principal investigator of two research projects on synthetic biology. He is the author of *The Worth of a Child* (1996) and coeditor, most recently, of *Trust and Integrity in Biomedical Research: The Case of Financial Conflicts of Interest* (2010).

Christopher J. Preston teaches philosophy at the University of Montana. He is the author most recently of *Saving Creation: Nature and Faith in the Life of Holmes Rolston III* (2009) and editor of *Engineering the Climate: The Ethics of Solar Radiation Management* (2012).

Ronald Sandler is an associate professor of philosophy and the director of the Ethics Institute at Northeastern University. He is author of *Character and Environment* (2007), *Nanotechnology: The Social and Ethical Issues* (2009), and *The Ethics of Species* (2012) and coeditor of *Environmental Justice and Environmentalism: The Social Justice Challenge to the Environmental Movement* (2007).

Index

Basic Bioethics

Arthur Caplan, editor

Books Acquired under the Editorship of Glenn McGee and Arthur Caplan

Peter A. Ubel, *Pricing Life: Why It's Time for Health Care Rationing*

Mark G. Kuczewski and Ronald Polansky, eds., *Bioethics: Ancient Themes in Contemporary Issues*

Suzanne Holland, Karen Lebacqz, and Laurie Zoloth, eds., *The Human Embryonic Stem Cell Debate: Science, Ethics, and Public Policy*

Gita Sen, Asha George, and Piroska Östlin, eds., *Engendering International Health: The Challenge of Equity* Carolyn McLeod, *Self-Trust and Reproductive Autonomy*

Lenny Moss, *What Genes* Can't *Do*

Jonathan D. Moreno, ed., *In the Wake of Terror: Medicine and Morality in a Time of Crisis*

Glenn McGee, ed., *Pragmatic Bioethics,* 2d edition

Timothy F. Murphy, *Case Studies in Biomedical Research Ethics*

Mark A. Rothstein, ed., *Genetics and Life Insurance: Medical Underwriting and Social Policy*

Kenneth A. Richman, *Ethics and the Metaphysics of Medicine: Reflections on Health and Beneficence*

David Lazer, ed., *DNA and the Criminal Justice System: The Technology of Justice*

Harold W. Baillie and Timothy K. Casey, eds., *Is Human Nature Obsolete? Genetics, Bioengineering, and the Future of the Human Condition*

Robert H. Blank and Janna C. Merrick, eds., *End-of-Life Decision Making: A Cross-National Study*

Norman L. Cantor, *Making Medical Decisions for the Profoundly Mentally Disabled*

Margrit Shildrick and Roxanne Mykitiuk, eds., *Ethics of the Body: Post-Conventional Challenges*

Alfred I. Tauber, *Patient Autonomy and the Ethics of Responsibility*

David H. Brendel, *Healing Psychiatry:Bridging the Science/Humanism Divide*

Jonathan Baron, *Against Bioethics*

Michael L. Gross, *Bioethics and Armed Conflict: Moral Dilemmas of Medicine and War*

Karen F. Greif and Jon F. Merz, *Current Controversies in the Biological Sciences: Case Studies of Policy Challenges from New Technologies*

Deborah Blizzard, *Looking Within: A Sociocultural Examination of Fetoscopy*

Ronald Cole-Turner, ed., *Design and Destiny: Jewish and Christian Perspectives on Human Germline Modification*

Holly Fernandez Lynch, *Conflicts of Conscience in Health Care: An Institutional Compromise*

Mark A. Bedau and Emily C. Parke, eds., *The Ethics of Protocells: Moral and Social Implications of Creating Life in the Laboratory*

Jonathan D. Moreno and Sam Berger, eds., *Progress in Bioethics: Science, Policy, and Politics*

Eric Racine, *Pragmatic Neuroethics: Improving Understanding and Treatment of the Mind-Brain*

Martha J. Farah, ed., *Neuroethics: An Introduction with Readings*

Jeremy R. Garrett, ed., *The Ethics of Animal Research: Exploring the Controversy*

Books Acquired under the Editorship of Arthur Caplan

Sheila Jasanoff, ed., *Reframing Rights: Bioconstitutionalism in the Genetic Age*

Christine Overall, *Why Have Children? The Ethical Debate*

Yechiel Michael Barilan, *Human Dignity, Human Rights, and Responsibility: The New Language of Global Bioethics and Bio-Law*

Tom Koch, *Thieves of Virtue: When Bioethics Stole Medicine*

Timothy F. Murphy, *Ethics, Sexual Orientation, and Choices about Children*

Daniel Callahan, *In Search of the Good: A Life in Bioethics*

Robert H. Blank, *Intervention in the Brain: Politics and Policy*

Gregory E. Kaebnick and Thomas H. Murray, *Artificial Life: Synthetic Biology and the Bounds of Nature*

Printed in the United States
by Baker & Taylor Publisher Services